THE PROFESSIONAL
QUEST FOR TRUTH

SUNY Series in Science,
Technology, and Society
Sal Restivo

THE PROFESSIONAL QUEST FOR TRUTH

A Social Theory of Science and Knowledge

STEPHAN FUCHS

State University
of New York
Press

Published by
State University of New York Press, Albany

For information, address State University of New York
Press, State University Plaza, Albany, N.Y., 12246

Library of Congress Cataloging-in-Publication Data

Fuchs, Stephan, 1958-
    The professional quest for truth : a social theory of science and knowledge
/ Stephan Fuchs.
        p.   cm.
    Includes bibliographical references and index.
    ISBN 0-7914-0923-6 (alk. paper) : $54.50. — ISBN 0-7914-0924-4
(pbk. : alk. paper) : $18.95
    1. Knowledge, Sociology of.   2. Science–Social aspects.
I. Title.
BD175.F834 1992
306.4'5—dc20   91-3095
                                                                        CIP

10 9 8 7 6 5 4 3 2 1

Meiner Mutter, Eva-Maria,
in aller Liebe
und Dankbarkeit zugeeignet

CONTENTS

LIST OF FIGURES

ACKNOWLEDGMENTS

I cannot possibly mention all the people to whom I am indebted for helping me develop the ideas in this book. Randall Collins first suggested looking at science as a particular form of work organization. His theory about the relationship between the social structures of scientific communities and their cognitive practices forms the backbone of much of my own thinking. Jonathan Turner restored my faith in a type of social theory that wants to explain why things are one way rather than another. Much of my understanding of theory emerged as a result of our debates on hermeneutics versus science. Jon also started it all back in windy and stormy Bremen, my first alma mater.

As the editor of the SUNY series on *Science, Technology and Society,* Sal Restivo helped me all along with that combination of personal support, editorial advice, and friendly collegiate criticism that all authors should be able to profit from. Amongst many other things, Sal pointed out that not all constructivism has to be idealistic. Rosalie Robertson, SUNY's editor, showed great patience in helping me prepare a publishable version of my manuscript.

My friends Charlie Case and Steven Patrick were always there for me when I needed them. Steven Patrick also prepared all the graphics, and he did so with his usual calm dedication. But nothing would have been possible without my wife Peggy.

PREFACE

As of the 1990s, the sociology of science is in flux. There is not even agreement as to what is its proper name: is it "the sociology of scientific knowledge" (SSK)? Is it "science studies," or perhaps "science and technology studies," as are named a number of new academic departments and programs? Is it an approach within the history of science, or a theme within the philosophy of science—to some practitioners of these disciplines, a disturbing upstart, to others a new resource? Or is it a branch of the deconstructionism whose main home is in literary theory, and which some see as the overriding movement of the late twentieth century? And finally, if it is any or all of these, how much connection remains to sociology as a distinct approach in research and theory? Are we witnessing the grand success of the sociology of science in interdisciplinary imperialism, or on the contrary are we watching it dissolve into a common property of reflexive criticism throughout the intellectual world?

Stephan Fuchs's work offers a unique way of understanding what is happening to the discipline and intellectual programs of today. The strength of his work is ultimately sociological, drawing on core areas in which sociologists have some of their strongest capabilities. These are the understanding of organizations and professions, the settings that structure people's actions and thoughts, set the trajectories of what is possible in their careers, and provide what permanence there is to the negotiated flux of social life. For Fuchs, scientific disciplines are invisible organizations, bigger organizations cutting across more concrete organizations such as the universities or laboratories in which particular scientists work. Sciences are organizations insofar as they are regular and repetitive networks of interaction that exhibit their own structure: denser in some places than others, turned inward toward certain common foci of attention. These invisible organizations have their own hierarchies of power; they control access to material resources such as lab equipment, jobs, journals, and book publishing. Through these networks flow the distinctive ideas and arguments that make up the cognitive side of a science.

What advantage is there in looking at sciences as invisible organizations? The great accomplishment of organizational theory has been to show the systematic causes and effects of differences among kinds of organizations. Organizations vary greatly in terms of the tasks their members perform, the technologies used, and the environments that surround them; what we call an organization is a way of linking together the local activities performed under such conditions so that there is a flow from one place to another. What happens locally certainly constrains what people can do; if we are talking about scientists, the local production of laboratory activities, or of scientific discourse, is constrained by the kinds of local conditions analyzed in organizational theory. The organization as a whole consists in the flows of products or ideas through a larger network. This has a shape of its own. In this sense we can say there is more to the construction of a science than what happens locally; at the same time this construction of the totality falls into the set of forms that make the varieties of organizations. If scientific disciplines are organizations, there is a good deal about them that is predictable by the conditions given in organization theory.

Fuchs's work builds on a number of earlier lines of development. There is the synthetic accomplishment of a theory of organizational structures, carried out especially by Joan Woodward, Charles Perrow, James D. Thompson, Paul Lawrence, and Jay Lorsch. The organizational model was applied with varying degrees of scope to the differences among scientific disciplines, by myself, and by Richard Whitley. Fuchs provides a refined version of the organizational theory of the sciences, developing it into a dynamic model, and extending Perrow's theory of the organizational production of "normal accidents" into a predictive theory of scientific controversies. The scope of Fuchs's theory also encompasses the school of microsociological research, the ultra-empirical studies of laboratory life and scientific discourse that have been made so prominent by Steve Woolgar, Bruno Latour, Karin Knorr-Cetina, Michael Mulkay, Harry Collins, and others. For Fuchs, these studies cannot be a dismissal of larger, macro-level structures of science, because any local laboratory activity exists as science only insofar as it is linked into a larger invisible organization. What is observed locally is part of one of the larger organizational patterns. When viewed from the point of view of organizational theory, the issues of task uncertainty, mutual dependence, and reputation are found on this micro level as well as in the large. For similar reasons, the epistemological issues of constructivism and relativism that have become so prominent in metatheoretical discussions are issues of particular kinds of organizational features. Philosophical issues in the sociology of science are not free-floating; just how much relativism and self-conscious constructivism emerges is itself the product of organizational conditions found in some kinds of networks, and not in others.

Fuchs shows that some intellectual fields are structured so as to emphasize hermeneutic interpretations, while others have the organizational structure that produces objectified knowledge. Fuchs ends up both with a general theory of the social determinants of scientific production, and with a challenge to the current factions in the sociology of ideas and sociology of culture generally. He turns the sociology of the current intellectual field back on itself; there is no stepping free of social structure merely by virtue of criticizing it. The criticism is only effective insofar as it also is a form of restructuring social organization. Hence Fuchs's concluding title (which is also a program): "Hermeneutics as Deprofessionalization."

Fuchs's work is not only a contribution to the sociological theory of science, although it is certainly that. It is also a statement of awareness that current epistemological and theoretical stances in this field are part of its organizational politics. To advocate a particular kind of epistemology, or particular intellectual contents, is a move favoring a particular kind of organization of our social networks. But given that, what kind of intellectual products we ourselves are able to enunciate depends on the kind of invisible organization in which we are operating, our own thoughts are also shaped by their social roots. If there is room to maneuver in today's intellectual field, it is because our surrounding organizational conditions themselves are in flux. To understand why this is so is the next challenge for a reflexive sociology of science. Fuchs challenges us to be reflexive not only in the abstract, but in the concrete, while recognizing the real constraints and opportunities we have to deal with.

Randall Collins
Department of Sociology
University of California, Riverside

CHAPTER 1

Toward a Theory of Scientific Organizations

Currently, the sociology of science is the fastest growing area in the sociology of knowledge. The relationship between the two has never been one of mutual support and recognition. Classical approaches in the sociology of knowledge generally held science to be exempt from sociological analysis. The Mannheimian approach to worldviews as weapons in the competitive struggles between social groups, the Marxist theory of ideology as "false consciousness," and phenomenological analyses of the mundane interpretive practices that actors engage in to construct everyday lifeworlds all maintained that there was a special rationality invested in science prohibiting sociological explanation. Mannheim claimed that the social substratum of science consisted of free-floating intellectuals who were not blinded by particular class interests; Marx was a firm believer in the objectivism of materialist science; and Schutz argued that the methodical pursuit of rational research freed science from the dogmatism of the natural attitude. The natural sciences particularly were assumed to follow neutral and objective methods for designing and testing theories that would eventually correspond to reality and were not influenced by external social factors. Even sociologists who stressed the historical peculiarity of modern Western science, such as Max Weber, argued that the rational core of science was nonsocial and that scientific knowledge could claim universal validity.

This privileged status of scientific knowledge reflects the sacred role science plays in the public discourse of modern society and culture. Ever since the Enlightenment equated science with societal progress and moral emancipation from tradition and superstition, science has come to be viewed as the paradigm for all rational practice. Science has replaced religion as the most authoritative

1

worldview, but shares with religion a charismatic remoteness from the profanities of everyday life and mundane reasoning. The label "scientific" lends special credibility and authority to knowledge claims and discursive practices, and so social groups try to mobilize science in support of their interests. Since the cultural capital invested in scientific knowledge is a highly valued commodity in the markets of public opinion and exchange, powerful social institutions, such as the professions, the state, political parties, or social movements, compete for the support of science as a way to gain legitimacy and acceptance. In a sense, science has become the religion of modernity; instrumental reason and technical objectivity are the core of its cult.

Realist epistemologies have systematized the worship of science as the model and metaphor for the sacred values of truth, impartiality, and objectivity. The eminently religious character of this worship is evident in the dualistic structure of epistemological realism. According to realism, there are "internal" and "external" domains in science (Hooker 1987:206ff.).[1] Realism intends to hold the internal and rational core of science exempt from sociological analysis. While the external domain consists of all those contingent social and psychological pressures that might influence variables such as researchers' choice of problems and institutional affiliation, the internal or rational domain of science is assumed to react to epistemic pressures only. That is, realist philosophies draw sharp distinctions between "internal" and "external" aspects of science, between "rational" and "social" factors, and between "contexts of discovery" and "contexts of justification." Traditional philosophy of science is a prescriptive model of how good science should operate, and good science reacts only to the epistemic pressures emanating from reality and logic. The distinctions between external and internal aspects of science resemble closely the distinctions religions draw between the sacred and the profane, and so I shall develop further Bloor's (1976) point that in worshipping the sacred side of science, philosophy of science actually worships our most fundamental and cherished social institutions.

As opposed to the traditional sociology of knowledge and philosophy of science, Mertonian sociology of science[2] does approach science as a mundane social institution with its own norms, reward and peer inspection systems, and institutionalized roles. However, due to its normativist and functionalist biases, the Mertonian framework is largely insensitive to the actual internal workings of science. To some extent, the Mertonian paradigm still subscribes to the philosophical distinctions between internal and external aspects of scientific production and concerns itself mostly with those normative pressures operating in scientific communities that institutionalize the extension of certified knowledge and keep the deviance arising from conflict and competition at bay through social control. The Mertonians have primarily been concerned with the behavioral and institutional aspects of *scientists*, not so much with the contents and everyday operation of *science*. In this sense, the Mertonian paradigm still holds the core of science exempt from sociological inspection (King 1971:15).[3]

The most innovative and stimulating research in the sociology of science has been done by the constructivist "sociology of scientific knowledge."[4] Two classical studies have carved out the central problematic of this approach: Ludwik Fleck's *Genesis and Development of a Scientific Fact* (1935/1979), and Thomas Kuhn's *The Structure of Scientific Revolutions* (1970).[5] The sociology of scientific knowledge addresses the actual internal workings and contents of science. It disenchants the quasi-religious illusions about rational science that traditional epistemology and, to a lesser extent, the Mertonian school have been cultivating for so long. The new sociology of science abandons the internal/external distinctions and approaches science as an ordinary social field where interacting and negotiating scientists construct natural reality in a surprisingly mundane fashion. The sociology of scientific knowledge opens the black box of scientific rationality and inspects the actual internal dynamics of science-in-the-making. Nothing mysterious and culturally unique is found to happen in science once Pandora's box is opened.

For the most part, the sociology of scientific knowledge has followed the interpretive and microinteractionist turn sociology underwent in the late sixties and seventies. Most notably, ethnomethodological analyses of interactions-in-contexts are frequently called upon to reconstruct the microdynamics of everyday scientific production. Social studies of science reveal that the ethnomethodological modes of producing social order are congruent with the modes of producing natural order. In fact, social and natural order coemerge. Following ethnomethodology, the sociology of scientific knowledge reveals that natural facticity, like social facticity, is socially accomplished and not given as an external reality that simply awaits "discovery." Facticity is always a cultural artifact. Consequently, the interactions and negotiations between scientists are shown to resemble very closely the practices of accounting for reality studied by ethnomethodologists in mundane settings. In fact, scientists at the workplace look more like Garfinkelian sense-makers than Parsonsian, Popperian, or Lakatosian rule-followers.

The influence of interpretive microsociology on social studies of science is also visible in the case study approach to science. Sociologists of scientific knowledge go into the laboratory as participant observers; they conduct in-depth interviews with scientists about incidents of controversial science; they analyze the indexical discursive practices scientists engage in to account for their actions; and they follow scientists in their attempts to enroll other agents in order to create strong networks of support. As a result, the evidence gathered by microsocial studies of science is often of a rather unsystematic and impressionistic kind. Thick descriptions of individual events and settings have largely driven out comparative studies of science as an organization. However, this interpretive microfocus was probably called for to overcome the excessive institutionalism of the Mertonian school and to start addressing the internal workings of science-in-the-making.

This innovative approach to science explains why the constructivist agenda receives prominent attention in this book. I shall draw rather extensively on this school in my discussions of the microdynamics of scientific construction because this is where the great strength of constructivism lies. Constructivism casts a light on the everyday processes of manufacturing scientific knowledge and hence is indispensable for any analysis claiming to uncover the internal workings and actual contents of science. With the constructivists I believe that realist philosophy misrepresents the actual reality of science, and like the constructivists I hold that science is "socially constructed."[6]

However, the constructivist agenda suffers from several severe shortcomings. Partly due to its orientation to interpretive microsociology, constructivism shares the astructural biases of actor-centered approaches to social structure (Hagendijk 1990).[7] The "sociological" is too often reduced to interactions between people. The overriding concern with the interpretive microproduction of natural reality has forgotten the structural and organizational conditions of scientific production. The focus on interactions, conversations, and texts has led the sociology of scientific knowledge to paint an overtly discursive and idealistic picture of science. Science is presented as a form of mundane sense making that employs ordinary interactive and conversational devices to manufacture socially accepted knowledge claims. What is left out in this rather harmonious and cozy account are the material and structural contexts of scientific production, such as the distribution of intellectual property, the level of concentration in the physical means of intellectual work, stratification, or disciplinary variations in the ways of doing research.[8]

The micro-macro problem reoccurs in social studies of science as their inability or unwillingness to offer a social theory of science. It turns out that social studies of science are not all that "social" and remain much closer to realist philosophy than its proponents realize. The philosophical "science is rational" is often simply replaced by the sociological "science is social."[9] Most case studies of individual events or laboratories have a purely descriptive and narrative status. Lacking a comparative perspective on the contexts of scientific production, social studies of science fail to lead up to an explanation of how science is socially constructed, and why this construction differs between, say, high-energy physics and social theory. In social studies of science, nothing can be explained because nothing is allowed to vary, or because interpretive descriptivism creates hypervariability without a common metric that would allow for comparisons. As a result, many case studies of science exhaust themselves in dogmatic reiterations of commonplaces: that realist epistemology misses actual scientific practice, that social factors must be taken into account, and that science is socially produced. The sociology of scientific knowledge is still preoccupied with epistemological critique and, more recently, with its own textual practices. This has prevented the field from developing its own Strong Program, that is, a social theory of scientific knowledge. To show that science is socially constructed is not the result but the beginning of any strong sociology of science, for the important task remains to explain why construction varies from science to science and over time.

The complete absence of any theoretical and comparative apparatus is also responsible for the unsplendid isolation of the field. Ironically, in pursuing the sociological dimensions of scientific work, the sociology of scientific knowledge has moved further and further away from the main body of sociology and, even more distressing, from the sociology of knowledge. The real test of any sociology of science that claims nothing special is happening in science and that science is ordinary social activity must be whether or not it is able to relate its findings to a general sociology of knowledge. But the present isolation of the field implicitly reinforces the special and extraordinary status of scientific knowledge celebrated by orthodox realism. A strong and truly social theory of science, however, regards science as a particular case of knowledge production in general and offers an explanation for the differences and similarities between mundane and scientific reasoning. Although the elaboration of such a general theory is beyond the scope of the present book, I hope to make some suggestions as to how such a theory might look.

The constructivist agenda, then, has opened up a sociological way of analyzing science-in-the-making, but has failed to develop a general social theory of science. I shall draw upon constructivism in my treatment of the microproduction of natural reality in chapters 2 and 3, but will go beyond constructivism in my attempt to design a comparative and explanatory theory of scientific production. It will be shown that constructivism does not describe the nature of science *per se*, but one form of scientific production. Constructivism can then be explained as a special case of the theory of scientific production.

There is, however, one brand of constructivism that is not widely received among mainstream constructivists, and which comes much closer to the theory developed here. This constructivism also insists that scientific knowledge is socially constructed, but it manages to avoid the philosophical pitfalls of relativism and the sociological shortcomings of interpretive idealism. Instead, "social construction" is seen more in Durkheimian and materialistic terms, which reveal science as a conflictual and stratified struggle over organizational and symbolic property (see Restivo and

Collins 1982; Collins and Restivo 1983a, 1983b). This more structural kind of constructivism takes the social-construction metaphor of scientific knowledge not to mean that all is interpretation, and that all interpretation is relative and contingent, but rather that science is much like politics and social conflict. This view is more aware of the organizational and material conditions of scientific work, and hence has guided the elaboration of my own theory.[10]

This theory, which I shall call the "theory of scientific organizations," is based upon two major traditions in social thought: the neo-Durkheimian sociology of groups and group cognitions, and the technological paradigm in organizational research. Science is approached as a particular work organization whose technologies and social structures determine the ways in which groups of scientists do their research. I believe this strategy will overcome the current isolation of the sociology of scientific knowledge and will prove useful in linking the sociology of science to the sociology of knowledge. At the same time, this theory will place the micro-production of natural reality in the wider material and organizational contexts of scientific communities, and so will attempt to bridge the micro-macro gap that separates the Mertonian paradigm from social studies of science.

There are two important works that have begun to develop an organizational theory of science, and that have influenced my own thinking a great deal: Randall Collins's *Conflict Sociology* (1975:470–523), and Richard Whitley's *The Intellectual and Social Organization of the Sciences* (1984). While both works go a long way toward developing such a theory, I believe both fall short of fully realizing its potential for explaining the cognitive *contents* of science. Despite these important efforts, there still is a widespread belief that explaining matters of content is the privilege and exclusive domain of interpretive microstudies. The equations between macro = institutional norms and micro = contents of science are misleading. Among other things, the theory of scientific organizations can explain why some fields produce solid facts while others engage in informal conversation. It explains why some sciences look more like literature, and why certain fields are rather cumulative and mature while others are very self-critical and discursive. This theory also explains why some fields constantly inspect their classical origins and philosophical foundations while others move self-assuredly on the path of increasing knowledge. It offers an explanation of why some sciences alternate between normal and revolutionary phases whereas others produce constant yet invisible innovations. The theory of scientific organizations can explain why some fields are very reflexive and pluralistic while others engage in a more empiricist style. And it explains this in just the same way as it would explain differences in the cognitive styles of mundane social groups and organizations. Just as in science, there are more "relativistic" social groups and organizations, such as universities, that allow for more internal debate and openness, and there are more "empiricist" groups and organizations, such as welfare bureaucracies, that are very prejudiced and dogmatic. I believe the strength of the theory of scientific organizations lies in this ability to explain a wide variety of mundane and scientific cognitive styles with the same conceptual apparatus.

To be sure, the term "contents of science" is understood here in the more general sense of "cognitive styles" and "discursive practices." I do not claim to be able to explain why scientists X and Y prefer theory A over B at laboratory Z in the year of C. And I am not sure whether any sociology of science has successfully explained matters of content in just this very detailed way. But the theory of scientific organizations does explain variations in the cognitive styles or mental habits of scientific fields. It explains why some knowledge systems are more formalized and standardized than others, or why some scientists advocate a rather authoritarian and prejudiced view of their work while others subscribe to a more tolerant and relativistic *conscience collective*.

The theory of scientific organizations intends to integrate the constructivist microperspective on science with the Mertonian emphasis on the institutional mesocontexts of scientific production. The sociology of scientific knowledge is certainly justified in attacking the functionalist and normativist biases of the Mertonian school, and it is also correct in objecting that these biases conceal the contents and actual internal workings of science. But I think this critique has gone too far in abandoning *any* notion of community organization in science in favor of a radically situationist and actor-centered microposition. I shall try to recover the Mertonian perspective on systems of peer inspection and approval without losing sight of the internal dynamics of science-in-the-making. We are then able to explain the covariations between forms of community organization and cognitive styles in science.

Due to its comparative and theoretical orientation, the theory of scientific organizations can also detect some surprising similarities and differences between science and other professions. The proof of the frequent declarations about the ordinary and mundane character of science lies in the ability to explain what scientific workstyles have in common with those of other professions. Such questions cannot even be posed within a framework that exclusively deals with individual labs or episodes of controversial science. Thus, I shall analyze the patterns of stratification across various professions; a topic which, incidentally, microsocial studies of science completely ignore. The theory of scientific organizations also explains, for example, why the workstyles of hospital-based medical specialists are closer to those of research front scientists than to medical generalists with a solo practice, and why current deconstructionism equates science and sociology with literary criticism and rhetoric. From this perspective, we can address the issue of why gentlemen-amateur physics differs from modern physics, and why social theorists have more in common with poets than, say, with experimental small-group researchers.

The theory of scientific organizations also fulfills the "reflexivity postulate" advanced by the Strong Program in social studies of science. While the radical sociology of scientific knowledge, still captured and fascinated by philosophical problems, interprets "reflexivity" as the abstract possibility of knowledge, I take this postulate conservatively to mean that a sociology of science should include a sociology of sociology.[11] Hence, I shall apply the organizational framework to the current state of sociology and argue that its immature and discursive character is not due to its youth or complexity but its fragmented control structures. In particular, I suggest that the debates between interpretivism and positivism, or science and hermeneutics, are not really debates over methodology but over organizational politics.

I realize that "theory" in the way it is done here is not all that fashionable these days. The postpositivist and deconstructionist movements have severely shaken our confidence in science, theory, and explanation. Literary criticism, hermeneutics, and rhetoric are the modest replacements for the grand occidental metanarratives of science and rationality. Hence, the theory of scientific organizations anticipates at least three major objections from constructivist practitioners. Since most sociologists of scientific knowledge subscribe to an interpretive or ethnographic case-study approach to science, there will be some criticism against theory *per se*. Interpretivism favors thick descriptions over general explanations, adopts members' perspectives rather than an outside observer's, and emphasizes local and historical uniqueness more than comparative analysis. Undoubtedly, such a perspective has its merits, for without detailed case studies we would have no access to the internal workings of science. However, without some kind of comparative yardstick, the sociology of scientific knowledge will end up adding case study upon case study, without any means of assessing their relative status and broader significance. The theory of scientific organizations does not, of course, intend to replace or even supersede interpretive case studies, but instead understands itself as a way of making sense of the huge amount of detailed findings that have already accumulated.

The second objection against the theory of scientific organizations comes from the recent network model of technoscientific artifacts.[12] This approach believes that a general sociological explanation of science is impossible or even undesirable. Instead, it recommends to follow scientists in their efforts to translate other agents' interests and enroll them in strong "actor-networks" of support. There is not a sociology on one side that explains a science on the other; there are only weak and strong associations made up of heterogeneous human and nonhuman forces. In this view, the sociologist should not try to explain science on the basis of some preexisting conceptual schema, such as social factors, that enter the game from outside. Rather, one should start with nothing and watch how technoscientific worlds are gradually built up through the work of the participating agents.

This is an intriguing argument, but I think it is premature to drop all attempts at a sociological explanation of science. In fact, the network perspective has developed its own generalizations, such as "enrollment" and "translation," that are now being applied to particular cases of science-building. The network concept itself is an eminently sociological one, and it would pose no conceptual difficulty whatever to admit nonhuman agents into such conventional sociological networks.[13] It seems to me that the dismissal of any sociological explanation of science is based on a narrow interpretation of the social which comprises only interactions between people. Due to their isolation from the main body of sociology and their interpretive microorientation, sociologists of scientific knowledge often have a very limited perception of the larger field. For example, the theory of scientific organizations not only includes interactions between people, but also such nonsocial determinants as technology, the material means of scientific production, levels of resource concentration, and the like. In any case, I think that the possibilities of a social explanation of science have by no means been exhausted, and it is therefore rather premature to drop such attempts altogether.

A third objection against the theory of scientific organizations will likely be raised by those practitioners who are very impressed by the postmodern deconstruction of Western metaphysics and its metanarratives of truth, presence, and rationality. In this view, the whole point of a sociology of scientific knowledge is that there is no such thing as an accurate representation of an external and objective reality. There is no meaningful distinction between the word and the world or between accounts and reality (Woolgar 1988a). Neither does natural science represent the reality of the physical world, or the sociology of scientific knowledge represent the actual reality of science. Hence, the attention shifts reflexively to the textual practices that create the appearance of such an independent reality. The sociology of scientific knowledge then turns into a special form of literary criticism that deconstructs its own narratives. The goal is to detect and avoid any rhetorical devices that create the dangerous illusion that the text is actually about something or, even worse, corresponds to some external reality. There is no such thing as a reality that could be represented by accurate descriptions, there are only texts that should reveal their own rhetorical tricks so as to allow for multiple readings and contingent interpretations. As a result, the "reflexive" branch in the sociology of scientific knowledge (Woolgar 1988b; Ashmore 1989) experiments with "New Literary Forms" such as play, dia-

logue, and irony that avoid any realist connotations and are structured multivocally to give voice to conflicting readings (Mulkay
1985). In this view, the worst thing to have is an empiricist monologue in which one author has the nerve to present her views as
actually being about something other than itself.

I believe that deconstructionist reflexivity is yet one more du-
bious outcome of remaining overimpressed with epistemological
problems such as truth, representation, and relativism.[14] Only if
we remain fascinated with the philosophical problem of truth will
our inability to solve it distract us from making points and statements about the world. Attention to the rhetorical and textual
practices of science and sociology is very useful, but only when
conducted as a discourse that is about something and has realist
ambitions.[15] But if we deconstruct our own textual practices as not
being about the *world*, we get stuck with our own *words* and
become obsessed, just like realist epistemology, with the purity of
our own discourse, instead of talking about science. Even novelists
do not routinely deconstruct their own stories but do everything
they can to draw readers into their narratives. I don't see any
reason to be more fictional than fiction.

It seems to me that deconstructionists want to throw out the
baby with the bathwater. I agree that there are no translocal foundations for knowledge, but practices simply don't need any foundations to be working well. The practice of parliamentary politics
requires no foundation in a social contract, and the practice of
procedural legislation needs no natural law doctrine. In the same
way, science can do, and actually does, without epistemology. But
to say that science needs and has no epistemological foundations
does not imply that we should do literary criticism and textual
deconstruction instead. If this is acceptable, then the rush away
from realism and explanation into literary criticism and rhetorical
deconstruction is itself not well founded. The antifoundationalist
pathos impresses only those who naively believed that practices
were indeed based on foundations. But physics proceeds largely
unimpressed by deconstruction. Once we drop this belief in the
necessity of foundations, deconstructionism is hardly more than a
tempest in the teapot of pure philosophy.

What, then, is the epistemic status of my own account? From
an abstract philosophical viewpoint, I do not claim any special

cognitive privileges for my analysis of science. From a philosophical, though not sociological,[16] perspective it is just another story about science that tries to muster all kinds of allies to present itself as a credible claim. I have chosen the empiricist monologue not because I believe in safe transcendental foundations for knowledge, but simply because I think it is a great way of making a point. As Latour (1988a) says, the problem of most texts, especially in our discipline, is that they are not believed enough, not that they are believed too much. Hence, I see no reason to deconstruct my own story through new literary devices that give voice to alternative interpretations and conflicting readings. *Others* will disagree and deconstruct and be reflexive anyway, so why should I do their job? In fact, it strikes me that the epistemic arrogance and omnipotence of reflexive and deconstructive accounts is even more offensive than the empiricist monologue. For the latter gives readers and critics at least a chance to speak for themselves, instead of silencing them by anticipating their responses in an allegedly multivocal discourse that is nevertheless constructed solipsistically.

CHAPTER OUTLINE

In the second chapter, I discern some common underlying themes in the sociology of scientific knowledge. Since the new field owes more to Kuhn and epistemological critique than to Marx, Mannheim, or Merton, I first deal with the philosophical background of social studies of science. Epistemological critique deconstructs the representational metaphor of true scientific knowledge corresponding to reality and, in this way, prepares a *sociological* analysis of science's internal workings. Realist epistemology believed that science reacted to epistemic pressures only, and showing that science reacts to *social* pressures precisely requires the critical conversion of epistemic into social pressures. This conversion is accomplished by epistemological critique.

Although epistemological critique was valuable in opening up a sociology of scientific knowledge, it now tends to stand in the way of a social theory of science. In particular, the endless debates on relativism and epistemic reflexivity have led the field to spend too much energy on irresolvable philosophical metaproblems. In fact, the recent experiments with New Literary Forms threaten to transform the field into a special branch of deconstructionist literary criticism. Relativism and reflexivity are still treated as *philosophical* issues, but once we stop doing philosophy, relativism turns out to be a pseudoproblem, and reflexivity simply means that a sociology of science is self-exemplifying.

The third chapter is especially addressed to nonspecialists in the science field. It will review the main empirical findings produced by social studies of science and put them into a comprehensive framework. This framework draws on Latour's network theory of fact production. As opposed to other constructivist agendas, Latour's analysis lends itself to comparative theory, despite his own reservations against theory and explanation. I shall argue that most of the rather disorganized and fragmented research results in social studies of science can be understood as leading up to a theory of fact production. Scientists use material and symbolic resources to induce other scientists to accept their statements as premises for further research, which gradually turns statements into facts. The important difference between mundane and scientific knowledge is that the latter can draw upon more and more powerful material and symbolic resources that raise the costs for objecting to scientific statements. I believe this is the structural reason for the authoritative status of science: there is too much capital invested in science to successfully challenge it.

The fourth chapter moves toward a general theory of scientific organizations. Microsocial studies of science are typically narrative and nonexplanatory case studies of individual events in science. They fail to lead up to a general theory of scientific production because nothing is allowed to vary. Discursive practices, the local interactions between scientists, and the "social factors" that close controversies between conflicting scientific groups are all implicitly treated as constants. Ironically, the claim to have discovered the social "nature" of science echoes the philosopher's claim to have revealed the rational "nature" of science. But if scientific practices are treated as constants, nothing can be explained because nothing is allowed to vary. There is no *a priori* reason to assume that all science is the same and that scientific practices do not change over time.

Once we treat scientific practices as variables rather than constants, however, we can explain the findings by social studies of science as resulting from the structural arrangements of scientific communities. The Mertonian mesofocus on patterns of community organization is recovered, but without its normativist and functionalist biases. The theory of scientific organizations explains some of the most significant findings in social studies of science, such as the idiosyncratic structure of local scientific production, the conversion of statements into facts, and the dynamics of scientific change. For each case it will be shown that constructivism does not describe the social nature of science *per se*, but one extreme pole of what should be treated as a continuum of scientific practice. That is, it is demonstrated how scientific practices covary with the social-structural arrangements of scientific communities.

To strengthen the case for a theory of scientific organizations, the fifth chapter reviews some classical studies in the technological tradition in organizational research. The theory of scientific organizations has the strong implication that the crucial organizational dynamics do not vary between scientific and nonscientific organizations, and this implication will be verified. Similar task technologies generate similar structures and practices, whether in science or other organizations. This chapter will also identify the crucial variables in all technological approaches to organization, and will suggest a comprehensive contingency model of organizational structure serving as the starting point for the theory of scientific organizations.

The sixth chapter puts the sciences into a comparative perspective. As a profession, science shares certain characteristics with other professions, such as medicine, law, and art. It will be shown how an organizational cum neo-Durkheimian theory of professional production explains the patterns of stratification in various professional fields. Such a theory also explains differences in the workstyles of various professions by underlying differences in the structural organization of professional groupings. In fact, this theory even explains why there is so much postmodern skepticism about the possibility of general theory in sociology. Postmodernism likens sociology and social theory to literature because these two professional work organizations have similarly loose and informal control structures, and they also institutionalize comparatively low levels of professional autonomy and resource concentration.

The seventh chapter introduces the fully developed theory of scientific organizations. This theory explains variations in the cognitive modes and workstyles between various scientific fields by underlying variations in the forms of organizational control over scientific production. This theory explains why some scientific fields are more scientific and mature than others, why some fields change more through innovation and cumulation rather than fragmentation or migration, and also explains why some scientific fields are organized more bureaucratically than others.

The final chapter applies the general theory of scientific organizations to one of the most persistent debates in sociology and social theory. This is the debate between "interpretive" and "normative" paradigms, between "qualitative" and "quantitative" research, or between science and hermeneutics. This debate is usually conceived of as an ontological and epistemological debate over the foundations of the social sciences, but I suggest that this debate is a conflict over organizational politics and structure. The interpretive paradigm, or hermeneutics, arises as a problem in organizations with low levels of reputational autonomy and professionalization of control. Hermeneutics is what we obtain if we democratize and decentralize the control structures of scientific work organizations. Such fields produce conversation rather than facts, they worry a lot about their metaphysical presuppositions, and they also engage in constant reinterpretations of their history and authoritative classical scriptures.

Unlike many works in the sociology of science, which are targeted more exclusively on a fairly narrow specialist audience, the present argument intends to bring the sociology of science back to the core concerns of our discipline. One of these core concerns has always been the relationship between social structure, material organization, and modes of cognition. Weber's comparative sociology of world religions, Marx's materialist theory of the mind, and Durkheim's investigations into social imagery are classical attempts at explaining thought sociologically. The theory of scientific organizations belongs more comfortably in this tradition than in the specialist sociology of scientific knowledge, which is generally more narrative, interpretive, and astructural. In this sense, the theory proposed here is naturally "cumulative." "Cumulation" does not mean the realist hope in gradually getting to know the

Truth, but the intention to identify core problems in a variety of areas, and addressing them with the same basic theoretical tools. In sociology, it is much more common to argue from a rather narrow perspective, and to make no persistent efforts at integrating one's own views with other approaches. This has contributed greatly to the current fragmentation in our field. The theory of scientific organizations suggests ways in which this fragmentation might be overcome. It does so not by rigidly prescribing one "central notion," but by offering a general model of knowledge production. Within this model, it is possible to see what such various fields as the sociology of knowledge, the sociology of professions, the sociology of science, organizational theory, and the sociology of sociology have in common.

The theory suggested here is really a social theory of knowledge, not just of science. Science is one special, if very powerful, case of knowledge production. Explaining science reveals the hegemonic alliance of power and knowledge (Aronowitz 1988:3–34; Fisher 1990), or how knowledge comes to be more-than-knowledge. As Hooker (1987:195–206) observes, science has become an extremely powerful force shaping every area of modern life with an authority that is largely unchallenged and justified by its allegedly superior rationality and objectivity. The sociology of science must question these dangerous illusions. They are dangerous because they allow power to hide behind knowledge and Truth. Science has disenchanted cosmology and religion only to enchant itself as the religion of pure and disinterested rationality. If we don't want to surrender to the "experts" and their science, it is essential that we understand the nature of science's power. This power is *social* in its origin, not epistemic. That is, it can be granted or withdrawn, for it is not based on the intrinsic superiority of science as cognition. This means that the differences between scientific and mundane knowledge are not epistemic but social. The power of science is the power of its organization and resources, not its method.

Underestimating the power of science is as dangerous as overestimating it, or attributing it to the pure forces of rationality. There is a strong tendency in some sociologies of science to reduce science to texts and rhetoric, and then to deconstruct the objectivist narratives of science. The enormous power of science as an organization is left out in this cozy picture. And the allegedly critical attitude toward science underlying this approach is as harmless as deconstruction itself, for it falsely regards science as nothing more than a special form of writing. The authority of science is not simply grounded in its texts, but rather in its organization. The critique of science as power is not greatly advanced by criticizing science as text only.

The important consequence is that science deserves no more and no different *type* of respect than other powerful organizations. People are often willing to surrender to the experts because they are in awe of the privileged rationality of their science. So we accept the experts telling us what to eat, who we are, how we should live, and how to raise our children. Foucault calls this totalitarian regime of Reason "biopower," Adorno and Horkheimer call it "instrumental reason," Habermas calls it the "colonization of the lifeworld." Once we stop being too impressed by the experts and their science, and once we realize that their power is simply that of their organization, we can begin to loosen their tight grips on our lives.

CHAPTER 2

The New Sociology of Science: Philosophical and Sociological Backgrounds

Despite their short history, microsocial studies of science have by now become established as a "normal" scientific research tradition. The development of the field reveals a now familiar pattern in the institutionalization of novel research specialties (Mullins 1973). A few early programmatic announcements by intellectual founders in the mid-seventies (Barnes 1974; Bloor 1976; Mulkay 1979) were followed by the emergence of local research centers (such as Bath and Edinburgh in the United Kingdom) and specialized publication outlets (e.g., the journal *Social Studies of Science* and the *Sociology of the Sciences Yearbook*). This institutional basis has allowed for the rapid growth of distinct "schools" and research clusters. A considerable body of detailed empirical studies of science has transformed the early programmatic statements into a "normal" research specialty, the "sociology of scientific knowledge,"[1] with its accepted routine definitions of subject matter, methods, and doctrines. Like all normal science, the field has developed its central paradigmatic dogma: scientific knowledge is held to be "socially constructed." Controversies with protagonists of an orthodox Mertonian sociology of science have helped to carve out the field's unique disciplinary stake.[2] Most importantly, social studies of science claim to explain sociologically the *contents* of scientific knowledge, and it is this claim that is held to distinguish the field rather sharply from the orthodox Mertonian paradigm with its emphasis on the institutional structures of scientific communities. Partly due to the increasing size of practitioners' networks, the field has even begun to differentiate into a number of loosely defined subspecialties: the ethnographic study of labora-

tory life, the semiotic and reflexive analysis of scientific and sociological discourse, the "relativist" study of contemporary scientific controversies, the interest paradigm, and the network model
of technoscientific artifacts.[3]

Despite obvious differences and debates among these subspecialties, certain common themes are clearly discernible. These
are, amongst others, epistemological critique, the issues of relativism and reflexivity, and the Strong Program.

EPISTEMOLOGICAL CRITIQUE

Clearly, the emergence of social studies of scientific knowledge
was, and continues to be, triggered more by epistemological critique than by Mertonian sociology of science or classical sociology
of knowledge. References to Merton and the Mertonians are rare,
whereas references to post-Kuhnian epistemology are com-
monplace. The critique of the philosophical "standard model of
science" (Mulkay 1979) firmly belongs to the interpretive repertoire of social students of science and still precedes many case
studies as their general rationale. The analysis of how science is
"socially constructed" is routinely presented as a more realistic
alternative to philosophical accounts portraying science as the rational and methodical search for universal laws that can be tested
against empirical reality.[4]

Following "postempiricist" philosophers and historians of science such as Kuhn (1970), Rorty (1979), or Feyerabend (1970), the
sociology of scientific knowledge intends to replace the normative
philosophical model of how science should proceed on its way
toward true knowledge with a realistic account of how scientists
actually act and interact to manufacture socially accepted knowledge claims.[5] The bottom line of epistemological critique deconstructs the representational metaphor of true scientific knowledge as a Mirror of Nature that has dominated Western philosophy
since the seventeenth century (Rorty 1979). The core of postempiricist philosophy asserts that it is social groups of knowledge
producers, not reality itself, that select "true accounts" and "adequate descriptions" of reality. As a contingent social practice, science lacks secure foundations in reality and rationality. True
knowledge does not correspond to reality since "corresponding to

reality," just like "deviance" or "mental retardation," refers to a contingent label scientists attach to those conventionally accepted practices and cognitions that nobody cares or dares to question anymore. In this way, epistemological critique has set the stage for the social study of science. The philosophical question, "How is valid knowledge possible?" turns into the sociological question, "How do social groups of scientists manage to construct the appearance of factlike knowledge?"

More specifically, social studies of science are often prefaced and justified by some version of the so-called "Duhem-Quine" hypothesis (see Hooker 1987:331ff.). Briefly, the underlying argument is that knowledge claims are always "underdetermined" by the available evidence, so that alternative claims are always equally justifiable in the light of that very same evidence. Hence, reality itself cannot decide which of the alternative interpretations is to be chosen, and this is where social factors—such as power and reputation—come into play. Since the "pressures of reality" are never strong enough to force scientists into accepting one, and only one, statement as *the* true representation, the "pressures of society" must account for choices between competing statements.

In its most radical and sophisticated Durkheimian version, this argument maintains that science and knowledge in general react to social pressures *only*, or that the epistemic pressures scientists experience from reality really are social pressures, disguised as the neutral and transcendental forces of Reason, language, and sensory perception. This argument suggests such a fundamental sociological gestalt-switch in our familiar understanding of cognition that it warrants brief discussion. For purposes of illustration, let me take an example from outside science: reports about mental states. I have chosen this example because it illustrates rather nicely what is meant by "reacting to social and epistemic pressures," and because it clarifies what is meant by converting epistemic pressures (philosophy) into social pressures (sociology).

When persons report about their mental states, such as pain or happiness, we usually do not call into question the truthfulness of such reports. If someone tells us that he or she is happy, and if we have no reason to doubt that person's honesty, we feel that we cannot disagree. Unlike reports about mental states, however, statements about external reality can be disagreed with, even if we are entirely convinced of a speaker's sincerity. We can be entirely certain that someone honestly believes in the truthfulness of what he or she is saying when asserting, "It is raining outside," while still disagreeing with that assertion. But we feel that it is impossible for us to disagree with someone we are convinced is honestly expressing pain or joy. Where does this asymmetry in reports about internal and external reality come from?

The standard philosophical answer is that this asymmetry results from ontological differences between two types of worlds and corresponding semantic differences between two types of statements (see Sellars 1963). Reports about mental states cannot in principle be disputed since they refer to an internal world of subjective feelings to which persons have privileged epistemic access. Reports about external reality, on the other hand, refer to a world that is equally accessible to all those who can see. According to the philosophical view, we feel an *epistemic* pressure not to disagree with reports about mental states. There is an intrinsic and noncontingent epistemic difference between reports about mental states and statements about external reality that forces us to not question the former while leaving room for disagreeing with the latter. Philosophy generally understands itself as the special science for such noncontingent epistemic pressures.

For the sociologist, matters are just the reverse. We refrain from disagreeing with mental reports not because individuals have privileged access to their inner sensations. Rather, privileged epistemic access to inner worlds *results* from the social convention not to question individual reports about mental states. The philosophical "privileged epistemic access" simply codifies the social habit to be polite and discreet when persons express their feelings. The mysterious philosophical dignity that surrounds reports about mental states expresses the social dignity of the modern self. Not questioning mental reports reflects our moral respect for personal integrity that is deeply entrenched in our social and cultural practices of worshipping the self as a sacred object (Goffman 1967). And not disputing reports about inner worlds is one way of acknowledging the sacred integrity of the self. Questioning such reports would amount to breaking the social taboo not to violate the privacy cult that surrounds the modern individual.

From a sociological perspective, then, we react to *social*, not to epistemic, pressures when realizing it is inappropriate to question authentic expressions of individual feelings. Or, more correctly, epistemic pressures—there is a group of inherently unquestionable statements about internal states that enjoy privileged access to reality—turn out to be social pressures: the modern self is worshipped as a sacred object whose privacy ought not to be interfered with. In Durkheimian terms, epistemic pressures are nothing but reified social pressures, for the epistemic force that seemingly drives us not to question mental reports reifies the social forces prohibiting intrusion into individual privacy.

That is, we can imagine societies without individual privacy cults that do not institutionalize privileged access to mental states and hence do not afford special epistemic status to reports about internal reality. But this is just another way of saying that such societies would engage in different ritual practices, worship different sacred objects that radiate different social pressures, and give privileged epistemic status to different kinds of statements.

For example, the Azande regard the statement, "the poison oracle never errs," as true by definition (Evans-Pritchard 1976). For the Azande, this statement cannot under any possible circumstances be called into question by contrary evidence. For us modernists, of course, there is no such thing as a poison oracle, and hence we regard the Azande belief as one of the superstitious and unscientific myths primitive minds are trapped in. But what the Azande belief really tells us is that conceptual truths—statements that cannot be refuted by contrary evidence—codify social and cultural practices so fundamental that they are being protected by epistemic immunity against criticism and alternative interpretations. The poison oracle is as fundamental to the Azande way of life as the sacred self is to ours. Consequently, in both societies, statements pertaining to these deep institutions are afforded a special epistemic status that protects them from being called into question. Such conceptually true statements are deemed true in all possible worlds and thus indicate a structural inability or unwillingness to imagine otherwise.

That is, conceptually true statements refer to social institutions so fundamental that they appear as natural, universal, and unchangeable. Privileged epistemic status *is* privileged social status. Generally speaking, by learning about a society's epistemology or about its epistemic distinctions between statements that cannot be false and statements that are open to criticism and revision, we learn what that society's most fundamental social and cultural practices are.

The above debates about mental reports and poison oracles very closely resemble the typical debates between philosophers and sociologists of scientific knowledge (see Phillips 1977:55ff.). The sociology of scientific knowledge reconverts the epistemic pressures philosophy says science is following into social pressures. For the realist philosopher, "good" science reacts exclusively to epistemic pressures, while social factors intrude into science from the outside only to distort its accounts of reality. Good science reacts to rational pressures (the transcendental rules of correct logical reasoning), to the pressures of language (statements that are true by definition), and to the pressures of reality itself (the empirical evidence gathered by sensory perception). Again, the strongest possible argument to be advanced by a sociology of scientific knowledge is to show how these epistemic pressures reify social pressures.

This is precisely Wittgenstein's (1967) strategy in his conventionalist sociology of language games (see Bloor 1983). For Wittgenstein, the epistemic force of our reasoning practices does not emerge from their privileged contact with reality but from the social authority invested in conventional habits of the mind, such as being able to correctly continue a series of numbers (2,4,6,8, . . .). It seems that continuing the series by following the rule "add 2" is natural and evident, for there is no other apparent way of "correctly" continuing the series but by 10,12,14, etc. The series seems to continue itself, generating each next number according to the epistemic pressures of a "self-evident" mathematical algorithm. But for Wittgenstein (1964), the ability to correctly continue a number series is due to training and routine practice, not to the epistemic necessities of mathematics. "Add 2" appears so obvious and natural because it is a rigidly trained habit. Fundamental social practices appear as transcendental epistemic necessities:

> Isn't it like this: so long as one thinks it can't be otherwise, one draws logical conclusions. This presumably means: *so long as such-and-*

such is not brought into question at all. The steps which are not brought into question are logical inferences. (Wittgenstein 1964:155)

It is not reality that imposes itself on our true accounts; rather, such statements as "corresponds to reality," "follows logically," and "is true by definition" are epistemological compliments we afford to mental behaviors that are deeply rooted in our habitual ways of making sense of the world. Similar points are made by Goodman (1955) and Quine (1953) in their discussions of logical inference and conceptual truths, and by Phillips (1977) and Bloor (1983) in their Wittgensteinian sociology of mathematics.

From a Wittgensteinian view, science is just another form of life and does not provide the rest of our culture with transcendental foundations. The special authority granted to scientific knowledge is social in origin, not epistemic. It is because of this social authority that we believe in science as systematically producing valid knowledge, not vice versa. There is no fundamental epistemological difference between science and the poison oracle. In both cases, social authority is prior to epistemic authority.

THE ISSUES OF RELATIVISM AND REFLEXIVITY

The motif of epistemological critique in social studies of science has greatly been nourished by the rationality debate in cultural anthropology (Wilson 1970; Hollis and Lukes 1982). This debate was informed by Winch's (1958) hermeneutic sociology of language games and by Evans-Pritchard's (1976) report of the Azande mythology. The main divide in this controversy revolves around the question of whether a scientific critique of primitive thought is legitimate or based on a category mistake. For relativists like Winch (1964), such a critique is illegitimate because it would involve imposing modern scientific criteria of truth and validity upon mythological worldviews that employ radically different criteria. Criticizing mythologies as "unscientific" and "inaccurate" is just as illegitimate as rejecting the moral standards of the Azande as "unethical" for the sole reason that they differ from our standards. Just as there are no transcendental and objective standards of morality that could serve as a neutral yardstick to evaluate the ethical value of historically varying norms, so there are no objective and

transcendental epistemic standards for evaluating the truth claims
of historically varying worldviews. In fact, we cannot even be sure
that tribal mythologies are not completely misunderstood from the
start when interpreted as "theories about objective reality" whose
intention is to "correspond to the facts."

In other words, the simple fact of cultural and historical variation might be reason enough to give up the search for universal
standards of truth, good taste, and moral justness. But whereas
sociologists readily accept that ethical viewpoints and aesthetic
tastes vary culturally and historically, they are much more hesitant
to admit that the same might be true for truth.

For rationalists such as Horton (1970) and Habermas (1984),
the scientific critique of primitive thought is perfectly legitimate
because there are context-independent and transcendental standards of rationality. Although these standards are "fully developed" only in modern worldviews, they are held to represent universal principles of correct cognition and provide a common and
unbiased yardstick for the scientific critique of "irrational" worldviews. Whatever these principles might be,[6] they do not just reflect
the idiosyncratic standards of a particular culture but reside in the
universal discourse of humankind. Typically, rationalists are also
evolutionists, claiming that worldviews can be arranged on a historical continuum of increasing rationality (Habermas 1979).

The very same divide separates rationalist philosophers from
relativist sociologists of science. According to orthodox philosophy, science embodies context-free standards of rationality because
it reacts to epistemic pressures only. Modern science is entitled to
claim universal validity because in principle, anyone who can think
logically, understand the meanings of words, and perceive the
world would arrive at the same conclusions. Therefore, science can
provide the rational foundations for all cultural activities, and this
ability explains science's privileged status as a special system of
knowledge.

The sociologist of scientific knowledge, however, points out
that science reacts to social pressures—just like any other ordinary
and mundane system of action. In this view, science is like art or
politics and cannot claim special epistemic status (Bourdieu 1975).
Typically, this argument draws upon some radical interpretation of
Kuhn's (1970) "paradigm incommensurability" hypothesis (Col-

lins and Pinch 1982). Like tribal and modern societies, competing scientific paradigms, in periods of revolutionary breakthroughs, represent incommensurable "forms of life" that lack a common frame of reference and hence cannot be evaluated rationally. Since there are no universal standards of rationality, science cannot serve as the general model for how all our cultural discourses should be organized.

In other words, most social students of science subscribe to some form of "epistemological relativism." Steve Woolgar (1983) conveniently distinguishes between three prevalent epistemological positions: reflective, mediative, and constitutive. The reflective position is that of traditional realist philosophy. Reality is ontologically independent from cognition but can be accurately represented in true descriptions. This is possible because reality ultimately selects its own accurate accounts through the methodical arrangement of the empirical evidence.

Most sociologists of scientific knowledge adhere to the "mediative" epistemological position. In this view, our accounts are "underdetermined" by empirical reality and thus are at least partly influenced by social factors. Reality exists independently from our accounts, but does not fully determine them. That is, objective reality has *some* bearing on scientific knowledge, but defines only the boundaries of a vague interpretive space of alternative accounts that are all compatible with the available evidence. Which of these alternative accounts will be accepted as true knowledge depends on contingent social processes.

Finally, the "constitutive" position radically denies an independent external reality existing "out there" apart from our accounts. This position appears to describe the epistemology of approaches such as discourse analysis and reflexivism that regard science as a special form of literary production (see Mulkay 1985; Woolgar 1988b). Since we have no way of deciding whether statements correspond to reality except by means of other statements, it makes no sense to assume the independent existence of an external reality to begin with (Woolgar 1988a). What we call "facts" are skillful constructions that manage to conceal their selectivity and contingency. Facts are produced through a "splitting and inversion" process in which reality is constructed, but then appears as the independent cause for this construction (Latour and Woolgar

1986). Likewise, a discovery is not a sudden revelation of the true characteristics of a previously hidden fact, but rather a conflictual accomplishment that only later is redescribed as passive recognition. According to the constitutive position, reality is only available under some description, and this is taken to imply that accounts are the only reality there is: "The main conclusion is the constitutive point that the organization of discourse *is* the object (Woolgar 1988a:81)."

It is at this point that relativism turns into reflexivity. Philosophical critics of relativism usually propose some kind of *tu quoque* argument that intends to reveal the self-refuting and paradoxical consequences of relativism when applied to itself (see Ashmore 1989:87ff.). If all knowledge is relative, so is the statement that this is the case. If natural scientific knowledge is a contingent social construct, then the sociological descriptions of science also depend on social negotiations and cannot claim objective validity. Of course, on yet another metalevel, this latter statement is itself only relatively true, which restores objective validity only to be destroyed at the next metalevel, and so on.[7]

It is those paradoxical "strange loops" and "tangled hierarchies" that fascinate reflexivism, the *dernier cri* in the sociology of scientific knowledge. According to reflexivists such as Woolgar (1988c) and Ashmore (1989), the sociology of scientific knowledge is particularly prone to reflexive investigations because it constructs knowledge claims about how scientists construct knowledge claims, and so methods, styles of presentation, and subject matter are inextricably intertwined.[8] The sociology of scientific knowledge critically turns to itself and inspects its own ways of fabricating knowledge claims, of writing texts that appear to refer to an external reality, and of rhetorically manufacturing the realist illusion of representation. After having deconstructed the realist metaphor of scientific knowledge as a Mirror of Nature, reflexivism sets out to detect and deconstruct the realist self-(mis)understanding in social studies of science.

Malcolm Ashmore (1989) has most relentlessly prosecuted cases of realist nonreflexivity in the sociology of scientific knowledge. A prominent realist offender is Kuhn (1970) who claims realist status for his own historiographic accounts but relativistic status for revolutionary paradigmatic upheavals.[9] If applied to itself, Kuhnian historiography cannot be read Whiggishly as a more adequate account and as a progressive replacement of previous Whiggish historical descriptions. Instead, Kuhnian historiography is one more self-contained paradigm practiced as a way of doing history by a community of historians. Ashmore (1989:89ff.) shows that this reflexive reading of Kuhn may have two conflicting implications: either a Kuhnian reading of Kuhn suggests that Kuhn is wrong because his relativism is self-refuting, or Kuhnian historiography is self-exemplifying and confirms that there are only historically bounded paradigms—in science as well as in the history of science.

The prime target of Ashmore's (1989:112–38) antirealist crusade is Harry Collins (1975, 1981d) and his Empirical Program of Relativism (EPOR).[10] Collins tries to avoid the allegedly self-refuting implications of relativism by banning reflexivity and by claiming special status for his sociological accounts of science ("special relativism"). While he believes that natural science is a contingent social accomplishment, EPOR is paradoxically credited with realist dignity. Collins (1981a:216,n.2) flatly declares:

> My prescription is to treat the social world as real, and as something about which we can have sound data, whereas we should treat the natural world as something problematic—a social construct rather than something real (!).[11]

One manifestation of this "special relativism" is Collins's (1981b) astonishing claim that his replication studies have now been replicated by his own research and that of his colleagues. This claim is astonishing because Collins's original finding is that there is no such thing as independent replication in science, only social factors and informal negotiations that close scientific controversies. Ashmore, detecting flagrant nonreflexivity here, replicates Collins's replication claim and ends up with a paradox: if Collins's original claim (that there is no such thing as replication) is right, then his own further studies and those of his relativist colleagues are not replications either, and then he is right because his position is self-exemplifying. If his original claim is wrong (there is such a thing as replication), then his own studies have been replicated, and hence confirm his original claim. The same paradox arises on the level of Ashmore's metareplication: if Collins's replication claim turns out to be problematic and contingent, then Collins's original claim is justified (replication is always problematic), but if Ashmore indeed replicates Collins's replication studies, then Collins's position is self-contradictory.

I think Collins's relativism is paradoxical only if we remain impressed, like Ashmore, by the philosophical way of thinking about science. Instead of claiming special relativism and replication status for his own findings, Collins should just have said that his studies and those of his colleagues are simply self-exemplifying. A *paradox* (and Ashmore's book about it and other paradoxes in the sociology of scientific knowledge) emerges *only* if the old notions of truth, rationality, and replication are tacitly maintained. There is no danger whatever for Collins and EPOR in admitting that their own research has the same status as that of the scientists they study. And reflexivists find paradoxes only if they somehow retain the discredited metaphors of realist epistemologies. The methodological horror of relativism and its paradoxes scares and occupies only those who still search for safe epistemic foundations.

This is not to say, however, that there are not some good reasons to retain the *practices* of realist writing and discourse without their philosophical foundations.[12] I have said before that realism is an excellent way of making points and arguments, even if we drop the notion that there is such a thing as adequate representation. Reflexivists, however, want to abandon the realist mode altogether and avoid any representational connotations that would lead their texts to be interpreted as being about some external reality, and representing that reality more or less accurately. Reflexivists are afraid to make statements that appear to be about something, and so they are quick to deconstruct their own statements by pointing to the textual and rhetorical practices that create the illusion of objectivity and authoritative interpretations. Hence, they experiment with New Literary Forms that essentially create multivocal texts that give voice to conflicting interpretations and alternative readings. Reflexive texts always display their own authors in order to highlight the constructive and selective work that goes into writing them. They gather authors, readers, and critics in a multivocal discourse that deconstructs the empiricist monologue and its representational fictions. Texts must be multivocal because there is no such thing as authoritative interpretations and definitive readings.

The *merit* of reflexivism lies, I believe, in having demonstrated what happens when we apply the sociology of scientific knowledge to itself. It is indeed paradoxical and pointless to claim realist status for one's own discourse while denying such status to the discourse of natural science. Reflexivism casts a light on the textual practices that writers employ to make credible claims and plausible statements that appear to refer to some external reality. But ironically, we can learn this lesson from reflexivism *only* when we interpret its arguments *realistically* as being about something, such as textual practices and representational devices. Understanding a claim realistically means taking it seriously. A realistic claim requires for itself that it is correct and consistent, that it says something that has not been said before, that others have reason to accept it as well, and that it can defend itself against contradictory claims. What is so bad about this? There are only two possibilities: either reflexivism is about something, and then it cannot help being interpreted in a realist way, or it is about nothing, and then it is difficult to see why there should not rather be silence.[13]

In a realistic interpretation, reflexivism is about the textual practices and stylistic devices employed in writing science and sociology. This is a most legitimate field and has greatly enhanced our understanding of how texts assemble all kinds of allies and forces to induce the reader to accept the author's statements and to use them in her own work.[14] Realistic reflexivism is nothing but discourse analysis. Nonrealistic reflexivism remains stuck in the self-referential textuality of its own rhetorical practices. But we have to interrupt self-reference *somewhere* to avoid dealing with nothing but paradoxes and tautologies (Luhmann 1989). And it really does not matter where we interrupt self-reference, for, as Latour (1988a:169) says, there are no "hierarchies of reflexivity." This means that it does not really matter whether we are naive realists and make straightforward statements about the world, or whether we are not so-naive-discourse analysts and make statements about how other authors make statements about the world, or whether we are really sophisticated reflexivists and make statements about how discourse analysis creates the fiction that there is such a thing as an independent reality of texts out there about which one can make statements. In fact, reflexivism might even be more dishonest than crude realism because it suggests that being more reflexive somehow yields a truer story, or a story that tells the truth that there is no truth. The main problem with increasing reflexivity is that we move further and further away from the world (science) and become captured in the webs of our own words.[15] From there it is not so far to the obsession with semantic purity that haunted the logical positivists. Organizationally, of course, reflexivism changes nothing. There are still books to be written, famous scholars to be enrolled for forewords, reputations earned, positions to be filled, and *Discourse and Reflexivity Workshops* to be attended. Despite the dramatic verve with which reflexivism sometimes announces itself,[16] the organizational song remains the same.

The theory of scientific organizations addresses the issues of relativism and reflexivity in a naively empiricist way. The basic argument is that certain scientific and mundane social groups and organizations are more prone to relativism and reflexivity than others. There are highly bureaucratized and routinized organizations, such as welfare services or the DMV, which are extremely nonreflexive and authoritarian in their practices. Fundamentalist sects also tend not to question their own foundations and to shield their practices against reflexive criticism. And then there are more loosely coupled and pluralistic organizations, such as universities, that allow for a more multivocal internal discourse. The same can be said about more informal social groups such as writers' guilds and professional associations that tend to give members a fairly high amount of discretion and work autonomy. From this structural perspective, relativism and reflexivity describe the epistemic self-understanding of groups with comparatively low internal coherence, transparent social boundaries, and flexible organizational policies and practices. Conversely, a more authoritarian, prejudiced, and rigidly "empiricist" cognitive style emerges in groups that are highly routinized, establish strict inside/outside distinctions, and are intolerant of internal deviance and dissidence.

A similar argument can be made for scientific groups. There are fields, such as social theory or literary criticism, that are very loosely structured and provide their practitioners with considerable work autonomy. Such fields are not very methodological and scientific and engage in rather informal discursive practices. My contention is that such fields are more likely to be critical of themselves, reflect a lot on their own practices, and interpret their knowledge as social constructs rather than as neutral facts about the world. These fields do not command very powerful resources and allies, for their discourse is mostly textual and nonexperimental. A good example is the sociology of scientific knowledge which allows for a great deal of internal debate and controversy, and encourages individual originality and creativity more than adherence to central paradigmatic dogmas.

Conversely, there are fields that are much less relativistic and reflexive. In the social sciences, experimental small group research is a good example.[17] Such fields are more closely coupled and firmly integrated. As a result, they believe more in the objectivism of their practices and in the realism of their methods and discourse. There is a fair amount of agreement on central paradigmatic dogmas, and so practitioners believe in safe philosophical foundations and scientistic rigor. Practitioners have faith in the superiority of their methods and in the cumulative advances of knowledge toward truth.

Reflexivism in the sociology of scientific knowledge pays so much attention to textual practices and rhetorical devices because texts are all there is in weak disciplines. Weak disciplines do not have laboratories and survey machines and numbers and statistics. *They only have words, and this leads them to believe that the word is the world.* Since they can only rely on the power of their own words to convince fellow practitioners, weak disciplines emphasize rhetoric and stylistic devices. Their loose internal organization is more conducive to conversation than to fact production, and so they mistrust and deconstruct all manifestations of representational realism. There is not a lot of faith in objective knowledge and accurate representation because individual practitioners have so much discretion that they can perceive all knowledge claims as active constructions rather than passive representations.

Relativism and reflexivity, then, are not problems that all fields have to deal with to the same extent. In fact, some fields have to be *told* that their knowledge might be a social construct and that their facticity is a subtle textual illusion. For the most part, it is sociologists and historians of scientific knowledge, not the practitioners of science themselves, who draw attention to the textuality of science and to the issues of relativism and reflexivity. These issues are not equally likely to emerge in all fields, and so they have to be introduced to some fields from the outside.

I shall interpret "reflexivism" for my own work to mean that a sociology of scientific knowledge should include a sociology of sociology. Sociological reflexivity differs from philosophical and textual reflexivity. Textual reflexivity transforms the field into literary criticism, deconstructs textual practices, and celebrates multivocal discourse in New Literary Forms. Textual reflexivity cancels the distinctions between fact and fiction, the word and the world, and remains in the realm of the knowing subject which is itself replaced by the "knowing text." Textual reflexivity creates the impression that texts are all there is. Sociological reflexivity, on the other hand, maintains the practice, though not the epistemology, of realist discourse. It makes statements about science and, reflexively, about sociology while leaving the deconstruction of these statements to others. Sociological reflexivity is not concerned with the paradoxes of relativism and reflexivity because such a concern signals that philosophical questions are still being asked. Put bluntly, sociological reflexivity acknowledges that relativism is not a problem for the Ku Klux Klan, and that coalminers do not worry much about reflexivity. Only a self-referential academic discourse can create the illusion that there are only texts that need deconstruction. For sociological reflexivity, textual reflexivity is an ideology in the strong sense of the term.

To be sure, this sociological reduction of relativism and reflexivity does not solve the problems of knowledge raised by them. My position is philosophically agnostic. The best we can do, I believe, is to say that some social and scientific groups are more likely to be aware of these problems than others.

THE STRONG PROGRAM: ENTERING THE BLACK BOX OF SCIENTIFIC RATIONALITY

David Bloor's *Knowledge and Social Imagery* (1976) is certainly the most important early programmatic statement announcing the new sociology of scientific knowledge. The core of Bloor's so-called "Strong Program" asserts that the very content of scientific knowledge is amenable to sociological analysis. This assertion sharply separates the sociology of scientific knowledge from classical sociology of knowledge and from Mertonian sociology of science. The classical program held scientific knowledge to be exempt from social conditioning, and the Mertonian paradigm never addressed the very contents of scientific knowledge claims. Of course, it is an open question whether the sociology of scientific knowledge has really met both requirements of the original Strong Program (Bloor 1976:1): to explain the *contents* of scientific knowledge *sociologically*. I shall come back to this issue on later occasions. But clearly, the Strong Program introduced a radically novel way of studying science. For the first time, it proposed to explain the very core of scientific activity.

Therefore, the unusually strong responses authors such as Kuhn (1970) or Bloor (1976) himself received from the (philosophical) scientific community are not at all surprising (see Lakatos and Musgrave 1970; Laudan 1981; Brown 1984). The intensity of these responses went far beyond routine academic critique, for they were triggered by vigorous moral outrage over the program to treat science just like any other profane social field and system of knowledge. In the orthodox common sense and philosophical views, there is something special and mysterious about science that prohibits "profane" attempts at explanation. This special and mysterious quality constitutes the internal, rational, and justification-related side of science, as opposed to the external, empirical, and discovery-related realm of profane forces. Philosophy specializes in the internal workings of science and reconstructs its rational structure, whereas sociology and history deal with the marginal external domain (Lakatos 1970; Laudan 1981; Hooker 1987).

The distinctions between "internal" and "external" factors, "rational" and "empirical" (sociological, historical, psychological) explanations, as well as between the two contexts of scientific activity exactly parallel the distinctions religion draws between the "profane" and "sacred" realms of society. As the religion of science, philosophy guards the sacred (rational, internal, justification-related) realm of science against polluting trespasses from those concerned with profane (empirical, external, discovery-related) matters. The religious task of philosophy is to preserve the sacred internal purity of rational science against the messy and vulgar distortions introduced into science by ordinary social reality. A powerful system of rituals such as Nobel Prize celebrations, scientific conferences, and public expositions of science maintains and emotionally reinforces these distinctions. Sociologists entering the internal workings of science do not just overextend their academic competences but break a taboo. Hence, orthodox realist philosophers regard a "sociology of scientific knowledge" as a provocative and outrageous absurdity, and they react to it with an honest and righteous sense of moral consternation. Is there a sociological explanation for this vehement reaction?

Following Durkheim's (1912/1954) sociology of religion, Bloor (1976:45) argues that the core of science's sacred status is society itself:

> Just as religious experience transmutes our experience of society, so on my hypothesis does philosophy, epistemology, or any general conceptions of knowledge.

Bloor substantiates this claim by showing how the debate between Popper and Kuhn over the nature of science parallels the controversies between Enlightenment and Romantic social ideologies. Popper's critical rationalism portrays science as the methodically controlled progress of knowledge within an egalitarian and open realm of free discourse, just like the Enlightenment's social and moral philosophy portrayed the evolutionary development of civil society as increasing democratic rationalization. Kuhn, on the other hand, describes science in the spirit of Romantic ideology, focusing on the dogmatism of traditions and the historically bounded rationality of conflicting paradigmatic communities. For Bloor (1976), this analogy between social ideologies and philosophies of science reveals the societal origins of science's sacred status—just like for Durkheim, the analogy between the power of the sacred and the power of society revealed the social origins of religion.

However, this analogy is problematic because Kuhn has contributed to disenchanting the sacred status of science. In fact, Kuhn (1970) describes (revolutionary) science as a profane political process. Kuhn's philosophy can hardly be interpreted as mystifying and worshipping science, although his analysis of scientific crises as resulting from perceived anomalies does have strong orthodox overtones, which brings him closer to Merton than many realize (Restivo 1983).

I suggest that in worshipping science as a sacred object, philosophy worships our most fundamental cultural practices. The binary oppositions philosophy establishes between "internal" versus "external" and "rational" versus "empirical" aspects of science correspond to the fundamentally dualistic structure of "occidental" culture. We draw deep ontological and epistemological distinctions between mind and body, spiritual and material essences, body and soul, subject and object, agency and behavior, intentions and causes, culture and nature, explanation and understanding. In each of these contrast pairs, two different types of forces are postulated: forces that are distinctly "rational," "human," or even "spiritual" (such as mind, agency, soul, and culture) versus profane and material forces (such as body, causes, and behavior). In fact, these binary oppositions express the ways in which occidental humans separate themselves as rational beings from that which is nonhuman and nonrational. Our cosmology revolves around the notion that the universe consists of two radically different types of things. These distinctions do not just concern the esoteric discourse of professional philosophers, but underlie our most basic cultural practices. They structure the ways in which humans appear to be different from animals (as rational and reflexive beings with minds, intentions, and culture), they structure our notions of moral responsibility (only intentional beings with agency can be guilty) and organize the intellectual cleavages in our academic system (*Geisteswissenschaften* versus *Naturwissenschaften*). The philosophical distinctions between the internal versus external and rational versus empirical aspects of science are congruent with these fundamental cultural dualisms. They separate the pure forces of Reason from the empirical circumstances of science and thus provide a general rationale for the basic oppositions in our culture.

In this sense, the sacred status of science is based on a conversion of cultural into philosophical distinctions, just as for Durkheim (1912/1954) the core of the sacred was society. The task of philosophy as modern religion is to prevent these distinctions that appear so crucial to our modes of experience from being blurred.

By separating internal from external aspects of science, philosophy acts as the religious guardian of rationality. In worshipping science, philosophy worships the purest form of rationality that has become the *Ersatzreligion* of modernity.

But there might be yet another, more specific sense in which philosophy can be said to worship society while worshipping science. In the orthodox view, there is an intimate link between science and democracy. Not only does "good" science require autonomy and independence from "external" political and economic pressures, but the scientific community itself is admired as the ideal type of a democratic community. In science, the only power operating is the rational force of the better argument. There may be stratification and rank differences in reputational statuses, but as "external" factors, these do not have a significant bearing on the internal workings of science. The scientific community institutionalizes norms of tolerance, skepticism, and intellectual communism that guarantee the free circulation of ideas within an egalitarian society of scholars (Merton 1973:267 ff.). The autonomy of science assures the rational progress of knowledge. A more or less unbiased system of peer inspection separates good from bad science, warranted from unwarranted claims, scientific from unscientific methods (Hagstrom 1965). Scientific socialization leads to the internalization of the "norms of science" so that external controls and formal sanctions can be kept at a minimum.

That is, in worshipping science, orthodox philosophy (and the sociology of science) also worships our most cherished political institutions. Orthodox philosophy of science is also political philosophy to the extent that science is held to institutionalize democratic values in their purest form. Whereas in profane systems of action, democratic institutions and citizen rights might occasionally and temporarily be suspended, science *is* freedom of speech and tolerance for dissident views. Without internal democracy, there would be no science, only ideology. In fact, early democratic social philosophy described emerging civil society as a large quasi-scientific community: citizens gathered in public places to freely and rationally discuss community affairs (Habermas 1962). Science is the archetype of "open society" (Popper 1957). The intimate relationship between democracy and science suggests that in worshipping the internal rationality of science as a sacred object, philosophy also celebrates the internal rationality of our democratic political institutions. The ethos of science is the ethos of freedom in its purest realization. The sacred status of science corresponds to the sacredness of our most fundamental political institutions.

Surprisingly, this reverential attitude towards science as democracy is even found in critical theories of society. For example, Habermas (1984) distinguishes between "communicative action" and "discourse." While communicative actions coordinate mundane activities on the basis of assumed interpersonal agreements, discourses are special communicative episodes geared towards reaching consensus. Unlike communicative actors, participants in discourses are suspended from the biases of the natural attitude and from ordinary pressures to act. Discourses institutionalize the idea of an "ideal speech situation" in which all participants have equal chances to make their contributions. Only the rational force of the better argument drives the discourse. Habermasian discourses, then, have the same status as the sacred side of science, for pure reason is nonsocial (Fuchs 1986).

It becomes now understandable why realist philosophers interpret the sociological "intrusion" into the internal workings of rational science as a provocative and outrageous attack on our most cherished social and cultural institutions. Attacking the internal core of science amounts to attacking our democratic political institutions:

> The clash between Popper and Kuhn is not about a mere technical point in epistemology. It concerns our central intellectual values, and has implications not only for theoretical physics but also for the underdeveloped social sciences and even for moral and political philosophy. If even in science there is no way of judging a theory but by assessing the number, faith, and vocal energy of its supporters, then this must be even more so for the social sciences: truth lies in power. Thus Kuhn's position vindicates, no doubt, unintentionally, the basic political *credo* of contemporary religious maniacs ('student revolutionaries'). (Lakatos 1970:93)

Following the general lines of reasoning suggested by Durkheim and Bloor, I have tried to point out the societal origins of science's sacred status. I have argued that in worshipping the internal and rational side of science, orthodox philosophy worships our most fundamental cultural practices and political institutions. The remaining problem is: how does philosophy accomplish this conversion of the social into the epistemic?

For Durkheim (1912/1954), religion accomplishes the conversion of society into the sacred through reification. The power of the sacred symbolizes the power of society, but the societal origins of the sacred become invisible. The authority of the sacred now seems to emanate from the sacred objects themselves, not from society. This process of conversion results in sacred objects and beings radiating a charismatic moral force that is worshipped as the transcendental and religious essence of Nature itself. Reification has transformed the pressures of society into the pressures of religious transcendence.

In similar ways, philosophy converts social pressures into epistemic forces. After philosophical conversion, scientific reasoning seems to follow the natural and self-evident laws of logic, language, and elementary perception. The pressures compelling cognition seem to originate in the natural rules of correct inference, in the meanings of our language, and in the nature of reality that imposes itself on our accounts. Philosophical conversion transforms and reifies social habits of the mind into transcendental rules of cognition:

> What, in the realm of language and ideas, we refer to as logical relations, and logical constraints, are really the constraints imposed upon us by other people. Logical necessity is a moral and social relation. (Bloor 1983:121)

Bloor's argument suggests a close analogy between the sacred and the logical as the core of rationality. Just as religion worships society in worshipping the sacred, so does philosophy worship society in celebrating the rational and internal side of science. If this analogy is valid, we should be able to discern in rationality the same formal properties that Durkheim (1912/1954) discerned in the sacred and that induced him to infer the societal origins of religion.

Durkheim's (1912/1954) strategy is to show how the formal properties of the sacred correspond to the formal properties of society as opposed to the individual. For him, this correspondence means that society is the origin of the sacred. Like society, the sacred has an essentially ambivalent character. First, sacred objects and beings radiate an *impersonal* force. Like society, the impersonal forces of the sacred command an individual's respect because they are perceived as superior to, and more powerful than, the individual. Like society, sacred forces are obeyed as *external* and independent forces that control individuals from the outside. But, at the same time, the external forces of the sacred reside *within* an individual as its soul or *mana*. The sacred resides both within and outside the individual, for it commands individual respect not as sheer external pressure but as a force that drives individuals from the inside.

Second, the sacred triggers ambivalent emotions that oscillate between respectful fear and devoted adoration. Like society, the sacred force is both attracting and repelling, desirable and dangerous, good and bad, rewarding and punishing. If approached improperly, the sacred turns against an individual, but if worshipped respectfully, the sacred is a source of strength and comfort. Approaching the sacred requires a special and pure attitude that brackets all profane and idiosyncratic concerns and interests. The impersonal force of the sacred commands purity of mind and innocence of intention and does not tolerate concerns other than worship for its own sake. All profane instrumental and idiosyncratic interests must be suspended lest the sacred be polluted.

The ambivalent character of the sacred corresponds to the ambivalent character of logic and rationality. Logical rules reside outside the individual and seem to represent universal and transcendental properties of correct reasoning. Like the sacred, logic is an impersonal force that compels cognition as an external and powerful authority. Logic commands respect, for there is no way an individual can arbitrarily interpret and modify its rules. A conclusion either follows logically from its premises or it does not follow at all. It cannot follow a little bit, under certain circumstances (other than those explicitly stated in the antecedent conditions), it cannot follow today but not tomorrow. It is not so much that our reasoning actively and deliberately employs logical rules; rather, it is *driven* by the force of logic. At the same time, however, logic resides *within* individual minds and is not a purely external force. Human beings have a competence for rational thought. The intelligent mind is the rational equivalent to the soul.

Moreover, logic and rationality, just like the sacred, trigger ambivalent emotions and attitudes. The forces of logic are cold, sober, and sometimes even merciless, but they are also enlightening, bright, and helpful. Logic is a powerful argumentative tool, for it is almost impossible not to accept a conclusion that "follows logically." Like the sacred, logic and rationality—the internal side of science—require purity of mind. The pursuit of truth must not be polluted by idiosyncratic interests and profane concerns. In the context of justification, profane and external concerns must be suspended, since only the pure forces of Reason are allowed to drive scientific inquiry. The scientific attitude must be impartial and objective and must not be influenced by nonscientific political preferences or economic pressures.

The scientific counterpart to breaking the religious taboos that prevent the sacred from being polluted by the profane is the *fraud* as the intentional distortion or manipulation of the evidence in the interest of "worldly" concerns. The scientific fraud arouses emotions similar to pollutions of the sacred, for it is an outrageous and severely sanctioned crime against the ethos of scientific rationality and impartiality. The taboos surrounding the fraud prevent the internal workings of science from being polluted by external and worldly influences. A scientist committing fraud can only be expelled from the scientific community so as to restore its collective purity, just like the breaker of a religious taboo must be banned from his tribe.

In short, there is a structural homology between the sacred and the logical and rational; and with Durkheim (1912/1954) I interpret this homology to mean that logic and rationality have social origins. Therefore, it is not surprising that the sociology of scientific knowledge received the same hostile reaction from philosophy that the sociology of the sacred received from religion. By entering the black box of internal scientific rationality, the sociology of scientific knowledge trespasses into the sacred and breaks the philosophical taboo. Philosophy reacts not just with mere academic criticism but with vehement moral outrage.

SUMMARY AND CONCLUSION

My goal in this chapter was to discern common themes in the sociology of scientific knowledge. Social studies of science define themselves not so much in opposition to classical sociology of knowledge or orthodox Mertonian sociology of science, but in opposition to traditional realist philosophy. Kuhn is considered more important for the field than Mannheim and Merton. The sociology of scientific knowledge started out as the sociological appendix to the postempiricist deconstruction of epistemology. The idealistic and rationalistic philosophical illusions about science had to be replaced by a realistic and empirically based sociology of scientific knowledge. For the first time, the Strong Program proposed that the internal workings of science were amenable to sociological scrutiny. Since science is our modern *Ersatzreligion*, philosophy reacted with moral outrage and open hostility. As the religious guardian of science and rationality, philosophy worshipped the epistemic purity of scientific method, protected the sacred core of science from polluting intrusions of the profane, and celebrated science as the rational foundation of our culture. Philosophy maintained that there was something very special and mysterious about science that escaped profane attempts at explanation. Science could explain anything but itself.

While epistemological critique was crucial for opening the new field, social studies of science have been overconcerned with philosophical problems, such as relativism and reflexivity. The prevailing philosophical way of asking questions about science has led the sociology of scientific knowledge away from the causal and comparative orientation of its own Strong Program. In its original formulation, the Strong Program was decidedly antiphilosophical. For example, the first tenet of causality called for a general theory of scientific knowledge production that could explain variations in epistemic styles across time and disciplinary fields:

> Such variation forms the starting point for the sociology of knowledge and constitutes its main problem. (Bloor 1976:3)

This strongly causal, theoretical, and comparative orientation has completely been lost in the sociology of scientific knowledge. To regain this orientation is one of my goals in developing an organizational sociology of scientific knowledge.

CHAPTER 3

Microsocial Studies of Science:
The Empirical Evidence

Social studies of science regard science generally as an ordinary social field of action without privileged epistemic and behavioral rationality. Nothing mysterious is revealed to happen in science once we open the black box of scientific rationality which philosophy has kept firmly closed as the sacred taboo of modernity. Inside the black box, we find mundane reasoners making sense, not rational reasoners discovering objective reality; we meet ordinary social actors engaged in everyday conversations, not quasiphilosophers following the rules of scientific method; we observe competitive owners of intellectual property fighting over credits and priority claims, not polite and disinterested participants in rational discourse committed to the collective search for truth.

In the present chapter, I shall review the empirical work in the sociology of scientific knowledge and suggest that Latour's (1987) theory of fact production is a good framework for integrating the rather fragmented research results.[1] Such integration is the first step toward cumulative theory. Facts are produced, not discovered; and just how this is done is one of the core questions in the sociology of scientific knowledge.

FACTS

The sociology of scientific knowledge regards the facticity of objective knowledge as a social accomplishment, not as an inherent characteristic that statements naturally display when they correspond to reality. Like the facticity of social order, the facticity of natural order is not given in reality itself but must be established and negotiated by members' social practices. Scientific facts are skillfully constructed "cultural objects" (Garfinkel, Lynch, and

45

Livingston 1981) that are always in danger of being reconverted into mere statements. Social and natural facts are fairly stable constructions, but *as* constructions, they exist only as long as no one gives voice to powerful and persistent disagreement. No established fact is immune against deconstruction. This is why science has a history.

Statements turn into facts through the "deletion of modalities" (Latour and Woolgar 1986) that surround particular statements. Modalities qualify the conditions under which statements can be suggested, interpreted, and possibly accepted or rejected. Modal operators such as those identifying time, context, and authorship of statements locate them in the actual here and now of conversations or textual discourse. Semantically, facts are statements without modalities. Facts have no visible authors and appear timeless and universal. But statements are always attributable to individuals or groups of authors. They are located at the intersection of subjective and objective worlds. They might turn out to represent nothing but an individual's idiosyncratic beliefs, they might be some "mixture" between subjective beliefs and objective reality, or they might turn out to represent no subject at all but reality itself.

As opposed to facts, statements occur as events in time and disappear into noise if no one listens to them and takes them seriously. And that is precisely the most likely outcome, for most statements are never listened to. Statements always depend on particular contexts, they refer to limited sets of external events and are always restricted by the particular conditions under which they may claim to be valid.

Consider the following example:

> In a more recent study, Taveggia and Ziemba (1978) find a tendency for women employees to be more strongly attached to extrinsic work features than men, after controlling for job status; yet this study considered only jobs of white vs. blue-collar job status and perceived mobility opportunity as controls. (Neil and Snizek 1987:246)

Several modalities qualify the above statement. It is attributable to particular individuals (Taveggia and Ziemba), and it is located in time (1978) as a concrete event. The reported finding is described as a tendency, not as an established regularity. Tendencies are weak and fragile constructions, for they cautiously admit to the possibility of being entirely spurious. To describe the reported finding as a "tendency" is thus to add the modality that this finding might just be noise, not a clear signal. Moreover, the sentence starting with "yet this study . . . " adds more modalities qualifying the conditions under which the reported statement should be interpreted and possibly accepted or rejected. This sentence suggests that the "tendency" observed by the authors might just be due not to reality but to the limited research design chosen by the authors: their job categories might not be fine-grained enough. Adding modalities has the effect of pushing statements toward the subject-pole of the subject-reality continuum. If the reported finding describes a "tendency" only, and if that tendency might in addition be due to a flawed measurement design, then the chances are that the authors constructed an artifact. By adding modalities, statements are transformed into more subjective and idiosyncratic beliefs; by deleting modalities, statements are transformed into mirrors of reality. Artifacts are constructed, facts are revealed. Statements are spoken by subjects, facts let reality speak for itself.

Consider now a second pair of examples, "DNA has the structure of a double helix," and "the SEI measures social stratification." These statements have no modalities qualifying their validity claims. They lack authors, contextual specifications, and references to time. Statements of fact do not give voice to opinions held by particular individuals in the here and now of an event, but seem to let reality speak for itself. Factual statements have erased all traces of social construction: the Watsons, Cricks, and Duncans have become more or less invisible, and so have the controversies that once surrounded these statements before they became facts. Statements of fact do not occur in time and cease to be made by individual speakers, for they are made by reality itself. Latour and Woolgar (1986:177) call this transformation the "splitting and inversion" of statements:

> The statement becomes a split entity. On the one hand, it is a set of words which represents a statement about an object. On the other hand, it corresponds to an object in itself which takes on a life on its own.

Once a statement is transformed into a fact, it becomes very costly and difficult to deconstruct. However, the very fact that science has a history means that even well-established facts can always be reconverted into statements. The chances are that in one or two hundred years, DNA will no longer have the structure of a double helix (just like it did not have that shape before Watson and Crick's "discovery"), and the SEI will be of interest to historians of social thought only. Whiggish historians of science will then investigate the "social factors" that led scientists to mistake erroneous opinions for facts of nature. Philosophers will explain these "errors" by the distorting influence of social ideologies and by the intrusion of external interests into scientific method. But in any event, facts can always be deconstructed by re-adding the modalities of time, authorship, and context that distinguish facts from mere statements. Deconstruction illustrates further what turns statements into facts: the social support of other people.

Statements are turned into facts by other people (see Ravetz 1971:181ff.; Latour 1987). To become a fact, a statement must determine the conditions under which other statements made by other people are possible. That is, a scientific statement must be accepted by other scientists as the basis or starting point for their own work (Luhmann 1984). In this way, contingency is slowly transformed into certainty. The more other scientists use a statement as the premises on which to build their own statements, the more they turn that statement into an unproblematic black box and an unquestioned foundation for subsequent scientific work.[2] Scientists do this to the extent to which they are convinced that their own work depends on a statement. A statement that wants to become a fact must interest people and try to convince them that *this* statement is crucial for *their* work. But it would be wrong to assume that the support of just any group of people will accomplish such transformations. Certain groups of statements have privileged chances to become facts. To be visible and taken seriously as "scientific," a statement must be made by someone who is a member of the relevant professional community and has access to some material means of mental production (see Lankford 1981; Rothenberg 1981).[3] To become a member and to gain access, university courses have to be taken, examinations passed, and credentials obtained.

In other words, statements and their authors must be part of relevant professional communities or networks to be taken seriously, and they need the support of these professional communities and networks to be considered as reasonable candidates for facts.[4] Professionalization means exclusion: only a very small number of statements have reasonable chances to become facts. All those statements outside professional communities have virtually no chances whatever once professional control excludes the vast majority of statements. Professionalization also means self-reference: only the support of *other scientists* can turn statements into facts. How exactly does this social support process operate?

Consider the following two examples:

[A] *Using* the Burke-Tully method, a black-white ethnic identity dimension is developed and used to measure ethnic identity among a sample of college students. The nature of this identity dimension is discussed and its relation to the other self variables is investigated. (White and Burke 1987:310, my emphasis)

[B] This study *uses* ordinal regression analysis to examine the impact of gender on work values, after controlling for various organizational variables. (Neil and Snizek 1987:245, my emphasis)

In both examples, the authors are actively supporting the transformation of particular statements and research practices into unproblematic black boxes or facts. In example A, the black box to be closed is the "Burke-Tully method" of measuring social identities, while in example B, "ordinal regression analysis" is pushed toward factlike status. In both examples, the authors accomplish this transformation by *using* these statements and practices as premises on which to build their own work. Fact production is an eminently practical accomplishment, for it involves doing more than saying. It is the practical and technical manipulations performed on material and symbolic objects that create facticity.[5] The "black-white ethnic identity dimension" from excerpt A is not something that is already out there; it must be constructed by translating the metric of the author's observations into the metric of the Burke-Tully method. Just like regression analysis in excerpt B, the Burke-Tully method operates as a "center of translation" and as a (more or less) "obligatory point of passage" (Callon 1986): these two methods try to insert themselves into other research practices and translate them into their own metric so that they can claim: whenever you try to measure ethnic identity or estimate causal effects you have to pass through *my* way of doing these things.[6]

By accepting and using the Burke-Tully method and regression analysis as (unproblematic) premises for their own (problematic) claims, the authors give their *social support* to these practices and thereby contribute to transforming them into "facts of measurement and effect estimation." Whoever questions the adequacy of the Burke-Tully method or the soundness of regression analysis must *now* expect the opposition of White and Burke or Neil and Snizek *plus* the opposition of all those who have used these techniques as black boxes previously or will do so in the future.[7] That is, White and Burke and Neil and Snizek, by actively supporting the conversion of statements into facts, strengthen the opposition that anybody will face who wants to open and inspect these black boxes, who wants to deconstruct their factual status, and who wants to turn them not into facts but into *just another* technique for measuring social identity and for estimating causal effects.

This opposition will be the stronger the more the authors' *own* claims depend for *their* factual status on such black boxes remaining closed. By using the Burke-Tully method and ordinal regression analysis as black boxes, the authors seek to bestow some epistemic authority upon their own claims. The factual status of their own claims depends on the black boxes remaining closed since questioning the Burke-Tully method or regression analysis would *automatically* imply questioning *any* claim that is built upon them. This is why it is so difficult to deconstruct well-established facts: opening such firmly closed black boxes amounts to questioning all claims that are built upon them, and to mobilizing the resistance of all the scientists who have built their own statements utilizing these black boxes.

Facts or black boxes display varying degrees of closure. Some black boxes—the well-established and undisputed facts as the "foundations" of a field—are very firmly closed, while others are more open, admit more critical light into their interiors, and hence are easier to inspect. That is, the social support invested in keeping black boxes closed is stronger in some cases, weaker in others. The Burke-Tully method from excerpt A, for example, is a less closed box than ordinal regression analysis from excerpt B. The Burke-Tully method is not a very widely known and used method for operationalizing "ethnic identity." The method is used primarily in a particular subspecialty of social research (gender and ethnic relations), and hence, the support network building upon and reinforcing its factual status is not very strong. Consequently, the method is (still) attributable to visible individual authors (Burke and Tully), must usually be explained whenever it is used (see White and Burke 1987:316ff.) and thus retains its status as one particular statement competing with other alternatives suggested by different authors.

Ordinal regression analysis, on the other hand, represents a much more firmly closed black box. Whoever uses regression analysis does not regularly have to comment on its statistical and mathematical foundations, it is just *used*. Questioning regression analysis would amount to questioning the building block of quantitative social research *per se*, not just a particular measurement instrument used by a few social researchers specializing in a subfield of the discipline. That is, the support network keeping the black box of regression analysis closed is much stronger than the one surrounding the Burke-Tully method. Regression analysis now has invisible authors, and it is implemented in standard computer packages such as SPSS-X and SAS. Regression analysis is routinely employed in quantitative statistical research, and it can be performed rather simply by pressing the appropriate buttons on the keyboard of a computer terminal. And every time the buttons are pressed, every time regression analysis is performed, it is turned into more of a closed black box, with the social support network growing stronger with every study that uses regression analysis as its unproblematic foundation.

Of course, turning statements into black boxes is not an irreversible process. As for the Burke-Tully method, we could question Osgood's semantic differentials on which it is based; we could launch a Gadamerian attack against misunderstanding *Verstehen* as technology; we could thematize the problematic hermeneutic translations between lay and professional sociological categories; we could criticize method and measurement in conventional sociology. The problematizing of statements opens black boxes, casts a critical light on their internal workings and tacit assumptions, and thus pushes statements toward the subject-pole of the subject-object or subject-reality continuum. In the case of the Burke-Tully method, such deconstruction would be rather easy, for the box is not that firmly closed to begin with, and the surrounding support network is not that strong.

Opening the black box of regression analysis would be a much more difficult task, for virtually all quantitative statistical research in sociology is based upon it. Of course, deconstruction is possible in principle: we could question the routinely made but usually violated postulates of linearity and normality underlying the "general linear model"; we could make a case against the substantive significance of "significance tests"; we could question the implicit assumption that equal codings indeed represent equal respondents' meanings; and we could reconstruct the invisible decisions that turned "messy" into "clean" data. In other words, deconstructing facts is always possible, but possibly very costly; depending on how firmly the black boxes are closed, how strong the support networks building upon their factual status are, and on how much research work would automatically be called into question as a result of opening its foundational black boxes.

Consider now one final example (from Garfinkel 1967:42–43):

[C]

    s: Hi, Ray. How is your girlfriend feeling?

    e: What do you mean, "How is she feeling?" Do you mean physical or mental?

    s: I mean how is she feeling? What's the matter with you? (He looked peeved.)

    e: Nothing. Just explain a little clearer what do you mean?

    s: Skip it. How are your Med School applications coming?

    e: What do you mean, "How are they?"

    s: You know what I mean.

    e: I really don't.

    s: What's the matter with you? Are you sick?

The difference between excerpt C and excerpts A and B from my previous discussion is not that A and B, as excerpts from research articles, exemplify pure scientific rationality, whereas C is just an instance of ordinary mundane reasoning. The difference is that in A and B, the authors try to close a black box (the Burke-Tully method and ordinal regression analysis), whereas in C, "E" opens a black box. "S" relies on "E" keeping the black box closed; that is, filling in the tacit background understandings that make indexical talk intelligible, but "E" refuses to cooperate. "S" builds upon the black box that contains the meaning of "how is your girlfriend feeling" as a fact of ordinary language and social life. "E" opens this black box, for he questions the self-evident and unproblematic certainties of the natural attitude that make smooth and orderly interaction possible. But note that neither regression analysis nor the double helix nor the meaning of "how are you feeling" are unproblematic in themselves; they are *turned* into facts by accepting them as premises of scientific work and social life.

In short, social and natural order rest upon the same basic mechanism of building black boxes and keeping them closed. The facticities of social and natural order are not given but must actively be accomplished through the coupling or nesting of statements and practices in surrounding statements and practices. Natural and social order emerge when certain statements and practices are transformed into facts of social life and objective reality. And in both cases, such transformations occur when *other people* support and use these statements and practices as solid black boxes on which to build their own statements and practices. Other scientists must use the Burke-Tully method and regression analysis as the unquestioned foundation of their own research, and other mundane reasoners must accept that the meaning of "how are you feeling?" is sufficiently clear to engage in orderly interaction. If we open black boxes, deconstruct facts, and disturb the orderly flow of social life, however, we trigger conflicts and controversies.

CONTROVERSIES AND CLOSURES

In the sociology of scientific knowledge, the study of contemporary scientific controversies assumes a prominent strategic position. As a subspecialty of the field, the so-called "Empirical Program of Relativism" (EPOR) developed by Harry Collins and his group at the University of Bath concentrates on episodes of controversial science.[8] Partly, this specialization is due to the vehement debates triggered by Thomas Kuhn's *The Structure of Scientific Revolutions* (1970), for these debates focused on a radical interpretation of Kuhn's treatment of revolutionary science as a profane political process. As extreme forms of controversies, scientific revolutions seemed to reveal the social nature of science. They were not resolved through crucial experimental tests or rational consensus but through ordinary political mechanisms such as rhetoric, propaganda, and struggles over organizational power.

The Empirical Program of Relativism (EPOR) emerged as a synthesis of these three motifs: the incommensurability of paradigms in revolutionary science, the absence of transcultural standards of rationality, and the critique of orthodox realist epistemology. In Collins's (1981b:3) formulation, relativism means that "the natural world has a small or (!) nonexistent role in the construction of scientific knowledge." Empirically, EPOR focuses on contemporary scientific controversies, not on historical revolutions.[9] The data are mostly gathered in in-depth interviews with the leading protagonists of controversial scientific positions. Since generally only a few well-known scientific leaders have been found to actively participate in controversies, respondents are described as forming "core sets" (Collins 1981c) that perform replications and close debates. Typically, respondents are asked to give their accounts of the controversies they are engaged in, and in practice, these accounts are more or less taken at face value.

Case studies of controversies within "orthodox" science include the contemporary debates about gravitational radiation (H. Collins 1975, 1981d), the hidden variables problem in quantum mechanics (Harvey 1981), the alleged discovery of magnetic monopoles (Pickering 1981), the memory transfer experiments with planarian worms (Travis 1981), and the detection of solar neutrinos (Pinch 1981). The general empirical outcome of these studies is that scientific controversies are not resolved by unambiguous experimental tests and the superior rational force of the better argument but by "social negotiation."[10]

Controversies assume a strategic position in social studies of science because they reveal how science is actually *made*. As opposed to "ready-made" or uncontroversial science, controversies open black boxes and disturb routine practices. Once we move back in time to the controversies that preceded normal science, the static picture of ready-made science suddenly begins to move. Students of controversial science witness science-in-the-making. Closed black boxes open and reveal controversial statements competing with other statements for recognition and acceptance (Gieryn and Figert 1990). The authors and actors of science suddenly reappear, they are engaged in heated debates, passionately accusing each other of "dogmatism," "incompetence," and "political biases" (Gilbert and Mulkay 1984). Tacit knowledge turns into noisy claims made by individuals; some pieces of lab equipment and certain "exemplars" in textbooks disappear altogether. Competing scientists engage in fierce priority disputes, and they mobilize organizational resources to suppress their competitor's claims (Collins and Restivo 1983b). Facts are still in the process of production. They do not yet correspond to reality but to the possibly erroneous opinions held by particular individuals at particular universities and laboratories. The SEI does not yet measure social stratification but is a suggestion made by someone somewhere in the literature. Objective reality cannot yet be mobilized in support of a particular statement, for it is the very nature of reality itself that is so controversial. There is some "evidence" evoked in support of claims, but competing statements use the same evidence under a different interpretation to support contradictory claims.

Ready-made science is more rational, peaceful, and certain; controversial science is more noisy, conflictual, and ambiguous.[11] Since they accompany science-in-the-making, controversies reveal how natural order is socially constructed, how open boxes are gradually closed, and how statements are slowly transformed into facts. Controversies are naturally occurring breaching experiments, or, in Collins's (1983:95) words, "autogarfinkels" for scientific knowledge. Like the ethnomethodological breaching experiments, controversies disturb the orderly flow of scientific practice. They turn the unproblematic into the uncertain, objective facts into contingent claims, and tacit into personal knowledge. By disturbing the facticity of social and natural order, controversies render visible how these orders are produced. Since order is produced by mobilizing strong support networks that can withstand trials of strength, *scientific controversies are conflicts over the control of social and material support networks*. The strategy is to enroll as many powerful agents as possible to support one's own claims and, simultaneously, to cut off conflicting statements from *their* networks. If this strategy is successful, then *closures* will settle controversies and transform statements into facts.

Closures transform controversial into normal science.[12] After closure, an important conversion takes place: objective reality itself is now held to have been the neutral arbiter selecting accurate representations and separating true facts from mere statements. Once certain statements are turned into facts and certain experimental procedures turn into routine practices, the controversial origins of these statements and practices are forgotten. Textbooks now introduce students to "normal" and natural ways of doing science. Pieces of technical apparatus are now tacitly used without remembering the complex history of their fabrication. Historians of science now write "rational reconstructions" of scientific development that attribute closure to the rational forces of crucial experimental tests and superior logical arguments. The "erroneous" statements that once competed with the now established facts are explained as being due to the intrusion of external forces into science's internal workings. Being retrospectively labeled "erroneous," these statements stood temporarily in the way of Reason progressing toward Truth.

Of course, just like the evolutionist historian of society, the rational historian of science obtains a Whiggish perspective that was not available to those involved in the original controversies. For in these controversies, the nature of reality itself was at stake, and competing statements all claimed to be true. At the time, there simply was no room for a historian of science claiming privileged access to reality and knowing in advance how a controversy was going to be settled. The interpretation of closures as being due to the uncompromising forces of Reason and Reality can only be an *historical* interpretation, for at the time of controversy, the voices of Reason and Reality are many.

In the sociology of scientific knowledge, the mechanisms of conflict and closure have not yet been systematically researched.[13] The Empirical Program of Relativism has been preoccupied with debunking realist epistemology and rational reconstructions of scientific change; that is, with showing how controversies can *not* be settled. Only a few unsystematic observations concerning possible closure mechanisms are scattered throughout the literature. Suggestions include the intention of scientists to "confine their arguments within a limited range of socially accepted conceptualizations of the natural world (Pickering 1981:87)"; the "use of rhetorical and presentational devices (Collins 1981b:5)"; "the reputation of various experimenters (Harvey 1981:109)"; the selective reporting of results in professional journals (Travis 1981); and the questioning of scientists' competence and personal integrity (Pinch and Collins 1984). The general argument derived from these studies is that settling scientific controversies is not significantly different from closing political or moral debates. "Contingent social factors" determine closures *regularly*, not just in exceptional and spectacular cases of "politicized" science.

In the remainder of this chapter, I shall draw upon the network theory of technoscientific artifacts to examine the mechanisms of closing scientific controversies more systematically. This theory argues that the authoritative status of science rests upon the powerful networks it mobilizes to support its statements and constructs (Latour 1987, 1988b; Callon, Law, and Rip 1986; Bijker, Hughes, and Pinch 1987). To induce other scientists to transform statements into facts, scientists may draw upon heterogeneous alliances and a variety of material and symbolic resources. The sociological difference between professional weather reports based on scientific meteorology and lay weather reports based on accumulated personal experience is not that the former are true and rational, whereas the latter are false and superstitious. The difference is that meteorologists are able to ground their forecasts on powerful social, symbolic, and material resources not available to lay people, such as certified professional communities, computer-processed data, satellite pictures, and the like. Such resources are difficult to deconstruct, and so account for the privileged status of scientific knowledge. In science, these resources include textuality, laboratories, and, missed by the network model (but not by the Mertonians and the "structural" constructivists), material and reputational property.

TEXTUALITY

Textuality is one such special resource available to science. Whereas mundane reasoning occurs mainly in verbal conversations, scientific discourse is in addition textual, for it produces certain kinds of written documents. Ultimately, "talking science"

(Lynch 1985) must lead to published research articles if claims are to be recognized and work is to be rewarded. Hence, one of the most important topics in the sociology of scientific knowledge is the stylistic, rhetorical, and semantic *conversions* that take place once the "mundane" reasoning in the laboratory gives way to the "scientific" reasoning in the research article. Medawar's (1969) early observation that the published article actually *misrepresents* the process of scientific investigation has repeatedly been confirmed in studies confronting laboratory reasoning with textual discourse (see Knorr-Cetina 1981; Latour and Woolgar 1986; Lynch 1985). The published article presents scientific activity as if it had followed the rational canons of scientific methodology all along. In the published article, science stops talking and lets reality speak for itself. The mundane reasoners of the laboratory disappear and give way to the rational reasoners of public science. The contingent and uncertain course of laboratory activity suddenly appears as the simple and predictable implementation of standard methodological algorithms. Garfinkelian sense-makers turn into Parsonsian rule-followers, and speakers of ordinary language turn into spokespersons for objective reality.

In the sociology of scientific knowledge, the textuality of science has become the specialty of a separate research program: discourse analysis.[14] Discourse analysis has been applied to such diverse issues as the semiotic structures of research articles (Gusfield 1976; Gilbert 1976; Law and Williams 1982; Myers 1985); the rhetorical devices used in scientific lectures (Woolgar 1980); the epistolary debates between scientists (Mulkay 1985); scientists' discursive practices (Gilbert and Mulkay 1982, 1984); the analysis of scientific humor (Mulkay and Gilbert 1982); and the emergence of experimental reports in the seventeenth century (Bazerman 1988; Shapin and Schaffer 1985). These studies concern themselves with the special features of scientific textuality, but they do not examine textuality as a resource *per se*. Before discussing the evidence gathered by discourse analysis, I want to investigate the systematic role textuality in general, not just scientific textuality, plays as a resource for transforming statements into facts.

In actual conversations, the opinions and interpretations suggested are always identifiable as the opinions and interpretations of particular individuals. In conversation, be it between lay actors or scientists, an interpretation of some reality always occurs in the here and now of copresent interactants. In actual conversation, an interpretation is always identifiable as a string of words uttered by a particular person at a particular point in time in a particular social and geographic location, nothing more. When copresent scientists, for example, discuss the significance of some observation, their contributions are sentences occurring in time and vanishing immediately once they are spoken. Other copresent scientists can always immediately question an interpretation, and they can always immediately question the possible justifications given in support of these interpretations. That is, conversations, even those in "talking science," are essentially nonauthoritative. They occur as temporal episodes, they are preceded and followed by other conversations giving voice to different interpretations, and they happen as events while countless other conversational events happen simultaneously.

Once we turn to written words or texts, however, the scene changes dramatically. The author of the written word is not visible as a copresent interlocutor; he or she is not identifiable as a particular individual who happens to defend a particular statement in the concrete here and now of a conversation. A written statement is not spoken by someone, and hence seems to speak for itself. In Luhmann's (1976) terms, a written statement will be *trusted* more than a spoken statement, for readers are more easily convinced than listeners.[15] A written statement does not seem to express an opinion but the "way things are." Like the reality it describes, the written word is "out there," independent from observers. The very existential mode of the written word maps the objective reality it purports to represent.

That is, by means of their own objective existence, written words appear to be closer to reality than spoken words, for the latter are spoken by a fallible epistemic agency other than reality itself. Unlike the spoken word, the written word is not an event, it does not vanish once it is read, but remains in its place for subsequent readings and thus has a stable historical existence. Readers of a text cannot demand instant justification for a claim; instead, they are referred to previous and subsequent passages in the same text, to other texts, from there to yet more texts, and so on. Textuality is a self-referential semiotic system: signs point at other signs that point at yet other signs (de Saussure 1966). Spoken words also point at other spoken words, but the difference between written and spoken words is that spoken words *also* point to their visible speakers. Textuality permits authors to produce "clean drafts." Texts lack the uncertainties, inconsistencies, and imperfections of the spoken word. No Freudian slips that would point at the fallible epistemic mechanism of the individual mind occur in texts.

Textuality by itself is a resource scientists use to gain social support for their statements. And recall that without social support, statements can never turn into facts. But in addition to textuality, scientific texts are able to command some special forces and enroll some particularly powerful agents not available to, say, crime dramas and detective stories. These special forces may be textual or nontextual. In Law's (1986b) words, scientific texts juxtapose heterogeneous forces, and it is this juxtaposition that motivates or even drives readers to transform statements into facts by incorporating them in their own work.

TEXTUAL AGENTS

Research articles generally use the "empiricist repertoire" to account for scientific statements and practices (Gilbert and Mulkay 1984; Gusfield 1976; Woolgar 1980). The empiricist repertoire employs the "style of non-style" (Gusfield 1976:17): statements are couched in neutral, impersonal, and purely representational terms. Unlike ordinary speech acts, scientific statements do not perform any actions, they just describe "what is out there." Statements are phrased in the passive mode to minimize the role of active agency in the fabrication of scientific claims. In scientific texts, "it was found that," "log-linear regression analysis was employed," and "data were collected." The authors of scientific texts withdraw from the frontstage of scientific action to let reality speak for itself. The research process is described as the simple and straightforward implementation of methodological rules. The complex uncertainties, errors, and decisions involved in actual scientific practice become invisible. In scientific texts, results were "found", not produced, and data were "collected," not manufactured. If the proper methodological procedures are observed, researchers are *driven* toward the truth. Conclusions seem to follow from the neutral and unbiased inspection of reality itself so that "the data are more compatible with model A," and the "evidence supports the conclusion that hypothesis B is warranted."

By minimizing the role of active and constructive agency in fabricating scientific knowledge, the empiricist repertoire mobilizes Reality itself in support of particular statements. In this way, statements are pushed toward the object-pole of the subject-object continuum. The expulsion of authors from scientific texts is further accomplished by internal textual organization. Scientific texts are usually written in the standard format of the conventional research report. The conventional format of an article signals that it is "in the truth" and deserves recognition as a professional and technically competent contribution to knowledge (Law and Williams 1982; Myers 1985). Typically, research reports are arranged sequentially into title, author's name and affiliation, abstract, introduction, hypothesis statement, methods and data description, results and findings, and conclusion or discussion. As a semiotic sign, this standard organization signals that results were obtained in a straightforward way by following routine procedures and standard methodologies. The standard format makes it possible to "trust in the truth" of a research contribution (Luhmann 1976). The conclusions were not drawn prematurely or intuitively, but at the end of a linear and cumulative research act; implying that *anyone* who had followed the same format would have arrived at the same conclusions. In Gusfield's (1976:21) words, research articles establish an "equivalence ratio" between authors and readers: both parties passively witness the gradual unfolding of the truth. For it is reality itself, not scientists, that forces our reasoning into accurate representations and correct conclusions.

In sum, scientists use the textual agents of reality and rationality in support of their claims. In this way, other scientists are motivated to use these claims in their own work, for these claims correspond to reality, were generated by rational scientific methods, and are couched in conventional textual formats to signal epistemic legitimacy. However, textual agents alone are not strong enough to mobilize the social support of other scientists. The textual agents of reality and rationality are used in virtually all scientific texts and hence cannot account for the *selective* recognition of certain statements as candidates for facts. Therefore, to gain social support scientists also call upon nontextual agents.

NONTEXTUAL AGENTS

Nontextual agents are the external resources authors of scientific texts can draw upon to raise the costs for objections to their statements (see Latour and Woolgar 1986; Latour 1987). Once it becomes too costly to object, a statement can either be flatly ignored or somehow incorporated into the work of other scientists. In the latter case, a statement is gradually transformed into a black box or a fact. How do authors raise the costs for objecting to their statements?

Research articles appear in refereed journals, preferably "prestigious" journals that publish the core contributions of particular fields. Usually, journals are published by some professional association or organization. The more prestigious associations and organizations generally publish the more prestigious journals. Articles submitted for publication are sent to reviewers who suggest acceptance, revision, or rejection of the manuscript. Usually, the editor goes along with the majority decision, but casts the critical vote if reviews are not unanimous.

That is, even before an article begins the author is able to draw upon a rather impressive nontextual support network. The author's statements do not trust that their intrinsic plausibility will convince all those who are open-minded and have eyes to see. From the outset, scientific statements are backed up by the prestige of particular journals and professional associations, by the reputation of editors and editorial boards, by the authority of the reviewers as specialists in a given field, and by the scientific organization the author is affiliated with. Before even beginning to read an article, the reader is confronted with a powerful support network that would have to be disconnected from the author's statements should the reader decide to "deconstruct" these statements. At the very least, the critical reader would have to mobilize support networks of comparable strength to suggest counterstatements that could reasonably intend to compete with the author's statements.

As the article moves on, the reader faces a gradual increase in the size and strength of the supporting textual and nontextual networks. Through references, the text relates itself to other texts that reference other texts drawing upon yet more texts, and so on. Often, references to other texts are not made to support a specific claim by yet another specific claim. Rather, they are generalized warning signals advising readers that should they decide to question an author's claim, they would have to question the claims of entire groups and communities of other scholars as well.

In sociology, generalized references are typically made to the "classics," most notably the Holy Trinity of St. Marx, St. Weber, and St. Durkheim. In sociology, the classics operate as cybernetic shorthands or paradigm substitutes identifying established traditions and foundations of the discipline. References to the classics— plus to some six or seven decades of exegesis—bestow the charismatic epistemic dignity radiated by the Founding Fathers upon a text, and thus are particularly difficult to question.

As opposed to participants in conversation, readers of scientific texts no longer confront a particular individual's opinion raised as a transient event in the here and now of face-to-face interaction. Instead, readers find themselves confronted with "the findings," the "rules of scientific method," and "established bodies of research and literature." Of course, the reader can in principle go to the library and check all the references to other texts for consistency with the author's claims, then check the references made in these other texts, and so on. The reader is now literally, *physically* surrounded by literatures that are invoked to join in a harmonious chorus of mutually reinforcing anonymous voices. To be sure, readers could work their way through the jungle of the literature, but the simple yet crucial point is: *they usually don't*. As we shall see shortly, this "they usually don't" is absolutely crucial for transforming statements into facts.

The nontextual support network is further strengthened as the text moves on. Turning the page, the reader finds, say, regression tables arranging numbers in support of a statement. The costs for objecting to statements that are supported by such numbers now become very high indeed. In addition to journals, reviewers, professional associations, scientific organizations, editors, and communities of cited scholars, the reader now faces the principles of statistics plus their mathematical foundations, standardized SAS or SPSS-X computer packages plus the companies that manufacture them, and, most importantly, *all the other scientists who use numbers and regression analysis in their own work.* To continue questioning a statement requires opening more and more black boxes that are firmly closed by the research and support networks building upon them (Latour 1987).

Again, the reader could in principle demand justification for each number. But imagine the time and costs such persistent objections would involve. Assuming that readers are sufficiently intimidated by the principles of statistics and their mathematical foundations, they could still demand that the authors show them how they obtained these numbers, what computer runs were performed (rather than other possible runs), how the raw data were "cleaned up," why constructs were measured in this way and not another, how the reliabilities of these measures were estimated, how the sample was drawn, and so forth. Still not convinced, readers could now demand to see the questionnaire, the graduate students who conducted the interviews, the transcripts of the interviewers' training sessions, the reports of the fieldwork, and so on.

In short, the reader may insist to *replicate* the author's research act step-by-step. But even if the reader did replicate the author's study, it is most likely that such a replication would not settle the controversy triggered by the reader's obnoxious objections, but would be one more controversial issue (Collins 1985). To be sure, replications do occur sometimes, but they happen rarely. The most likely outcome is that no replication will ever be performed. Replications are costly, time-consuming, and not considered original and innovative contributions. Furthermore, to perform a replication means not being able to promote one's own research during that time.

In most cases, the strength of the textual and nontextual support networks surrounding scientific statements leaves readers with only two practicable options: to ignore the author's statements, or to somehow incorporate them into their own work. Flatly ignoring the author's statement is not a very likely reaction, since the very fact that the readers read the article in the first place indicates that the author's statement is closely related to and relevant for the readers' own research. Hence, readers are more likely to "trust in the truth" of the author's statement, and to somehow use it in their own work. That is, *readers are likely to include the author's work in their own references*. Depending on whether the author's statements confirm or contradict the readers' own work, readers will either use these statements in direct support of their own work ("following the Burke-Tully method," "employing regression analysis"), or will somehow qualify the author's statements ("although X reports that y, this finding might be due to a flawed research design, limited sample, weak measures, inappropriate statistical techniques, etc."). In the first case, readers contribute to closing a black box and producing a fact. In the second case, readers leave the box open, but do not bother to "replicate" the author's findings. Controversial issues of minor importance are downplayed, treated with agnostic indifference, expected to be resolved by "further research," and finally forgotten (Fuchs 1986).

In sum, scientific texts mobilize textual and nontextual agents to gain the social support of other scientists who transform statements into facts or black boxes. The stronger the network of agents supporting a claim, the higher the costs for objections, and the more likely it is that such a claim will be used by other scientists in their own work. Statements that cannot call upon strong textual and nontextual agents are very unlikely to turn into facts. Such statements are not "in the truth," no one trusts them, they do not appear in prestigious journals, they cannot call upon many other texts for support. This is why novel statements are so controversial and scientific revolutions so rare. *Novel statements cannot (yet) mobilize strong social support networks*. Novel statements cannot draw upon established traditions of research and routine technologies, they cannot reference many other texts, no research builds (yet) upon them. Besides their authors, nobody is really too disturbed when novel statements disappear, for no one's work is closely related to them. Therefore, it is rather easy and inexpensive to deconstruct novel knowledge. Novel statements may perfectly correspond to reality, but as long as there are no strong networks of textual and nontextual agents supporting them, novel statements remain extremely fragile and vulnerable constructions.

Laboratories

Laboratories are especially powerful nontextual agents supporting scientific statements. In the sociology of scientific knowledge, laboratory studies have become institutionalized as one of the field's subspecialties (see Knorr-Cetina 1981, 1983; Jagtenberg 1983; Lynch 1985; Latour 1980, 1987; Latour and Woolgar 1986; Garfinkel, Lynch, and Livingston 1981; Star 1983, 1985; Traweek 1988). The popularity of laboratory studies stems from the opportunity to observe *in situ* how scientific work is actually being carried out.[16] Whereas discourse analyses rely heavily on written accounts, and the study of controversies is based largely on interviews with scientists, ethnographies of laboratory life have more direct observational access to the actual production plants of science. Therefore, lab studies seem to offer the best way to fulfill the main postulates of the Strong Program: to examine the internal workings of science, and to explain how the content of knowledge claims is actually being manufactured and negotiated by interacting scientists.

The typical method chosen for these studies is in-depth participant observation of one particular laboratory. Often, the setting is approached from an "anthropological" angle. That is, assuming an essential "strangeness" in local laboratory cultures is held to serve as a methodological tool for disclosing the latent background assumptions and tacit routine practices that structure the production of scientific knowledge.[17] For unlike the study of scientific controversies, ethnographies of laboratory life usually deal with the less spectacular process of "normal" science as it is done on a day-to-day basis. Whereas episodes of controversial science operate as naturally occuring breaching experiments or "autogarfinkels" (Collins 1983) in that they explicitly problematize tacit certainties, lab studies witness the more orderly and routine activities of normal science. Hence, approaching the field from an anthropological perspective is expected to reveal that which is tacitly implied in scientific practice; just like the routines and certainties of tribal communities are invisible to their members but immediately obvious as "strange" to their Western anthropological observers.

In the present context, I shall discuss laboratories as special resources scientists can draw upon to gain the social support necessary to transform their statements into facts. Texts are limited, for they can only call upon other texts or other semiotic signs such as numbers. Once we turn to the actual production sites of science, however, we find that *scientific statements are anchored in working pieces of technical equipment*. Laboratories are behind the numbers, graphs, and figures displayed in scientific texts (Latour 1987). The textual and nontextual resources discussed so far make it fairly difficult and costly to continue objecting to a statement, but the costs for objections rise even more dramatically once statements are backed up by machines. In principle, readers of scientific texts can try to sever the links authors have established between their statements and "bodies of research and literature." Readers can, in principle, replicate an author's research act to see whether the numbers gathered in support of a statement "really" support that statement, or could just as well be interpreted so as to support contradictory statements. But how does one "deconstruct" a working piece of technical apparatus?[18]

Upon entry into their settings, ethnographers of laboratory life report that nature and reality are nowhere to be seen (Knorr-Cetina 1981; Latour and Woolgar 1986). Instead, the laboratory is filled with instruments, artificial objects, and created substances (Ravetz 1971). The experimental conditions under which laboratory tests are conducted do not replicate the conditions found in nature. Instead, objects and substances are exposed to extreme temperatures, purified, isolated, and combined with other substances to create "unnatural" reactions and "new" substances. That is, laboratory reality is artificially created reality, it is not the "immediately given and obvious" reality postulated by the philosopher of science.

The primordial reality produced in the laboratory consists of certain readings which are taken off particular pieces of technical apparatus, such as the graphs produced by a chart recorder. By themselves, these readings do not yet constitute information, they must be "ordered," "cleaned up," and "interpreted" to make sense.

According to Ravetz (1971), these operations constitute the "craft" character of scientific work. Handling the technical equipment, interpreting the primordial readings taken off the equipment, and transforming these readings into meaningful information requires knowledge of a highly tacit and personal sort. It is the practical experience gained from *doing* science that enables the trained researcher to perceive meaningful signals where the novice perceives only noise. Craft skills are especially required whenever the boundaries between signals and noise are blurred and shifting, as is the case with novel discoveries. To extract such signals from surrounding noise involves skills that cannot be taught by formal instructions but that must be acquired in the process of *in situ* participation in research activities (Ravetz 1971; Kuhn 1970).

To transform the primordial readings taken off the technical equipment into the graphs, tables, and figures that appear in research reports, scientists use a variety of "inscription devices," a term Latour and Woolgar (1986) borrowed from Gaston Bachelard. The role of these devices in producing the artificial objects of science is the thematic core of laboratory studies. Most generally, inscription devices are the instruments scientists use to render nature and reality visible (Lynch 1985). For the reality appearing in laboratories and in the final exhibits presented in research reports is an artificially constructed reality, it is not a reality that can simply be observed. No one can simply observe chromosomes and income distributions: these are artificially constructed objects. In Latour and Woolgar's (1986:51) words,

> an inscription device is any item of apparatus or particular configuration of such items which can transform a material substance into a figure or a diagram which is directly usable by one of the members of the office space.

Scientists are constantly concerned with "inscribing," that is, with recording and classifying observations because *these inscriptions are the only reality available in the laboratory*. The task is for scientists to arrange the signals produced by the equipment into a coherent textual narrative which relates signs to other signs to make a "statement." Out of the chaotic mass of inscriptions scientists must create a coherent text, the research report, which presents the research act as the straightforward revelation of the facts of nature.

This task is greatly facilitated by the routine working of the technical apparatus or the inscription devices. These devices operate with a fair amount of predictability, just like the machinery used in ordinary material production. Due to the regularity with which certain inscriptions are obtained, the technical apparatus appears as a neutral instrument that permits nature to present itself to scientists. The inscriptions produced seem to represent nature itself, while the skillful craft operations necessary to extract signals from noise become invisible. The final inscriptions used to support statements in research reports seem to merely arrange the findings and the evidence in a convenient format. In other words, the technical apparatus *reifies* the constructive operations involved in fact fabrication so that results can be "found," data can be "collected," and reality can be "observed."

The ethnography of laboratory life, however, insists that reification conceals the essential *constructive* role the technical apparatus plays in producing the artificial realities of science (see Knorr-Cetina 1981, 1983; Traweek 1988:72; Galison 1987:251). In this view, the working apparatus consists of closed black boxes that once were open when the corresponding statements were controversial. That is, the technical apparatus has materialized statements that once competed with other statements for being enclosed in factual black boxes. And since closures of controversies and the production of black boxes are to a certain extent contingent events (meaning that other outcomes could have been obtained), the implication is that the technical apparatus is not just a neutral instrument giving voice to reality but thoroughly *constitutes* the artificial reality available in scientific inscriptions:

> The artificial reality, which participants describe in terms of an objective entity, has in fact been constructed by the use of inscription devices. Such a reality, which Bachelard terms the 'phenomenotechnique', takes on the appearance of a phenomenon by virtue of its construction through material techniques. (Latour and Woolgar 1986:64)

Let me illustrate this important point by means of a simple analogy. If we take the production metaphor employed by the ethnography of laboratory life literally, the production of artificial scientific inscriptions (graphs, tables, diagrams, etc.) is conceptually identical with, say, the production of cars. In both production processes, some raw material (a natural object or substance, steel) is transformed into a final product (a table or diagram in a research report, a car) through a series of constructive operations. Now, to say that the technical apparatus used in the laboratory is just a neutral device for revealing the "true nature of reality" makes as much sense as saying that the machinery necessary to produce cars reveals the "true car nature" of steel. That is, different types of technical equipment generate different types of products and outcomes, so that in science the inscription devices or technical instruments used *constitute* the artificial objects scientists construct to anchor their statements in "reality."

To use a different example from social research, the inscriptions presented in final research reports (i.e., the tables, graphs, and diagrams) appear as findings obtained from the neutral and unbiased inspection of reality. But the instruments researchers use to obtain these findings also *shape* the results; a point that has nourished the critical attacks from qualitative and interpretive approaches against mainstream conceptualizations of research (Cicourel 1964). For example, as an inscription device a questionnaire or interview schedule is not just a neutral instrument to solicit responses that correspond to the true internal states and external conditions of those interviewed. Rather, the shape of the instrument, such as the sequence of items, the wording of questions, and the response formats also *constitute* the very responses obtained. That is, a different instrument would yield different responses so that the "reality" presented in analyses of survey data is a thoroughly fabricated, not "observed" reality.

In any case, the technical apparatus used in the laboratory to produce the artificial inscriptions underlying the exhibits presented in research reports dramatically increases the costs for objecting scientific statements (Latour 1987). Readers are now confronted not only with self-referential bodies of literature, professional organizations, journals, and numbers but in addition with *working* pieces of technical apparatus. In Law's (1986a:39) words, "the laboratory is simply a site where a heterogeneous range of resources from both near and far are brought together and assembled into a hopefully coherent whole." Readers now face routinely used black boxes such as electron microscopes, the Stanford linear accelerator, and statistical software packages. Backed up by the technical apparatus, statements become even stronger in their efforts to force themselves upon the statements of other scientists. Laboratories are powerful tools to motivate others to transform one's statements into facts because it is extremely difficult to question something that obviously "works" and does so fairly reliably and predictably.

Property

Scientists draw upon a variety of textual and nontextual resources to induce other scientists to transform their statements into facts by using them as premises for their own research. So far, I have discussed resources that are more or less equally distributed among scientists. All scientists call upon the textual agents of Reality and Rationality, all scientists relate their statements to statements made by other authors, and to do science, all scientists must have access to some laboratory equipment.[19] But there is one resource that is not equally distributed among scientists: property. To conclude this chapter, I shall briefly discuss one symbolic (reputation) and one material (control over the means of mental production) property resource.

Reputation. To get their statements transformed into facts, other scientists must use these statements in their own work, such as in "following the Burke-Tully method," and "employing regression analysis." To influence other scientists' work, statements must first and foremost be *recognized*. Science is a noisy marketplace for statements competing for recognition. All statements demand being listened to. In this noisy and messy competition for attention, statements search for devices that make them better audible and hence more recognizable. This situation resembles a chaotic political rally in which everybody demands attention at the same time, and suddenly someone seizes a microphone. The person with the microphone has almost already won the struggle to control the rally because this person's voice is the only one everybody can hear.

Reputation is such a microphone. Statements made by scientists with high reputations have better visibility and hence are more likely to be recognized by other scientists. As a form of "particularized cultural capital" (Collins 1988:361), reputation gives some advance authority and credibility to statements. No scientist can listen to all statements. Reputation reduces complexity for scientists whose limited span of attention necessitates decisions about whom to listen to, whom to ignore, whom to ridicule, and whom to take very seriously. Reputation helps scientists make these decisions, and thus makes science a simpler world to deal with. Of course, this is not to say that statements suggested by scientists who enjoy a high reputation will always outcompete statements proposed by lower ranked scientists. But reputation does improve the visibility of statements and increases the chances that a statement will be recognized at all. In science, just as in any other social field, not all voices carry equally far, and reputation is one more resource scientists can use to give their statements more credibility, and thus to induce other scientists to use their statements. That is, scientists with high prestige are more likely to produce facts than scientists with low prestige.

In Mertonian sociology of science, this phenomenon is noticed under the labels "accumulation of advantage" and "Matthew effect" (see Zuckerman 1988:531–32). Scientists accumulate advantages such as resources and reputation, for present evaluations and future opportunities depend on past accomplishments. Hence, advantages accumulate at an accelerating rate. Especially those scientists who manage to acquire reputation and high status at an early stage in their career are later more able to capitalize disproportionately on their achievements. In a similar way, the Matthew effect works in favor of those scientists who already are widely recognized and visible, so that their work is more noticed and rewarded than the comparable contributions of their lesser-known peers. Together, accumulated advantages and the Matthew effect assure that reputational property increases like economic capital and, just like economic capital, creates sharp inequalities in status.

Like *mana* and charisma, reputation is, to a certain extent, transferable through personal networks. Working and being affiliated with famous scientists increases one's own chances of being recognized. Zuckerman (1977) observes that students of Nobelists are more likely to win that award themselves, and Collins (1989) documents how intellectual creativity and eminence are transmitted through master-pupil ties in philosophy. This process further increases stratification in science, and sharply separates core groups from those working on the periphery of a given field.

Control over the Means of Mental Production. Presumably, reputation correlates strongly with control over the material means of scientific production. The process of accumulating advantages ensures that symbolic profits are converted into material payoffs, so that scientists with high prestige and reputation are likely to control organizational property. The material means include experimental equipment, jobs, journal space, and the like. Whoever controls access to these means, the "gatekeepers" in science, can be expected to have considerable influence over the ways in which science is done. But, at the same time, this influence will be tempered by the high discretion scientists must be granted when dealing with uncertain problems.

Forms of control covary with the ways in which material and organizational property is distributed. Collins and Restivo (1983b) have shown that before the institutionalization of university-based professional research, scientific politicians used their control of property in a rather ruthless way to suppress opposing views and approaches. When scientific property is controlled by individuals or elite national Academies, powerful intellectual politicians will try to establish cognitive monopolies, and the overall mode of scientific work will be rather patrimonial and authoritarian. Control over property is extremely concentrated, practitioner networks are rigidly stratified, and the core group of powerful organization builders engages in very aggressive intellectual politics.

With the advent of university-based scientific communities, control over property becomes less individualistic and patrimonial. Modern science is a somewhat more open and collective enterprise. Property is controlled by organizations rather than individuals, and the systems of peer review and inspection endorse a more collectivistic and egalitarian workstyle. Individual contributions need the approval of the group and cannot earn reputations by simple authoritarian *fiat* or sheer oppression of alternative views. For the case of modern mathematics, Collins and Restivo (1983b:220) observe that

> collectivist attitudes among twentieth-century mathematicians have been structurally induced. Mathematicians have had to become altruists in order to pursue any major intellectual ambitions. The growth of the mathematical community and the development of numerous special fields threatened to make it difficult or nearly impossible for individual mathematicians to have their publications recognized.

This does not mean, however, that stratification of control and organizational inequality are absent in modern science. It is just that the modes of scientific production, communication, and recognition change once a patrimonial system of private property is replaced by a collective system of organizational property. But science remains highly stratified, and control over the material means of scientific production continues to influence the ways in which science is done. Especially in highly stratified fields with rare, expensive, and thus highly concentrated technical equipment—such as high energy physics—research directors will have considerable influence. Traweek (1988:5) notes that an important source of power in experimental particle physics stems from controlling access to "beamtime," that is, to accelerator beam pulses. But it is important to keep in mind that such control will be balanced by the collective organization of modern science, and by the autonomy of scientific workers.

Control over property and scientific work, then, will take place in more subtle forms, not so much through direct command and supervision.[20] Ethnographies of laboratory life (Knorr-Cetina 1981) inform us that laboratories establish distinct local "cultures" that tacitly prescribe certain ways of doing research. To the insider, certain labs are quickly identifiable in terms of their directors, research emphases, and peculiar ways of doing science. Laboratories develop reputations and gain credits for their local research culture. Laboratories are the organizational centers of particular schools and networks of teachers and students. Having worked in a particular laboratory identifies a researcher as someone interested in particular problems and approaching them from a specific angle.

The important mechanism here is that anyone who wants to perform a certain type of research is likely to pass through the local laboratory in which that research is being done. And the more concentrated the resources, and the more stratified the field, the more influence local cultures will have. In this way, laboratories become "obligatory points of passage" that scientists must pass through to do their work (Callon 1987; Callon, Law, and Rip 1986). Laboratories translate the research practices of their workers into local customs of doing research. By controlling local lab cultures, research directors are able to influence the *premises* of the research being done at a particular location. Most importantly, research directors do not have to dictate the terms of other scientists' research but can confidently rely on the local culture transmitting characteristic ways of doing science. By controlling local cultures and access to the material means of scientific production, research directors have privileged chances to determine what kind of research will be done at a given location, how certain problems will be approached, and what exemplary solutions will be chosen.

SUMMARY AND CONCLUSION

To say that science is a "social construct" does not simply involve documenting how a particular finding corresponds to a researcher's political interests, how a debate was settled by the *fiat* of a powerful scientist, or how editorial policies and funding agencies influence the ways in which science is done. There is a strong tendency in social studies of science, most notably in the sociology of controversies, to interpret the term "social" to mean that the task is to show how social factors influence the internal workings of science. But in this interpretation, the "social nature of science" reinstalls the old epistemological dualisms between "internal" and "external" aspects of science.

In the present chapter, we have seen that science is "social" at the very core of its most cherished and worshipped building blocks: facts. Statements are turned into facts by gaining control over the social support networks of science and over other scientists' work. Natural order coemerges with social order. To induce other scientists to turn statements into facts, researchers draw upon a variety of textual (reality and rationality) and nontextual (references, numbers, laboratories, property) resources. Facts emerge when these resources make it too difficult and costly to deconstruct the statements supported by them. Black boxes are firmly closed when a considerable body of research builds upon statements. The epistemic authority of facts is grounded in the social authority of scientific support networks. The difference between science and other (e.g., literary) forms of knowledge is not that the former represents the truth about objective reality, whereas the latter are works of pure fiction and imagination. The difference between science and all other forms of reasoning and knowledge is that science commands more and stronger resources to support its statements. Hence, scientific knowledge appears to have special epistemic authority, but it is the special social authority of science's superior resources that lies at the core of its privileged epistemic status.

CHAPTER 4

How Social are Social Studies of Science?

The greatest merit of social studies of science is that for the first time, Pandora's black box has been opened in order to subject the internal workings and contents of science to sociological scrutiny. The sociology of scientific knowledge has disenchanted our quasi-religious illusions about science as a culturally privileged system of purely rational and disinterested cognitive activities. Due to its apparent remoteness from mundane and practical concerns, science is the hardest possible case for demonstrating the social core of knowledge. The sociology of scientific knowledge has entered a cognitive domain that neither classical sociology of knowledge nor Mertonian sociology of science regarded as being amenable to sociological explanation.

In a sense, the sociology of scientific knowledge relates to science like Durkheimian sociology of religion relates to divine transcendence. David Bloor (1976) pointed to this analogy early on in the development of the novel field. Religion and science share the mysterious auras of cultural fields that must not be polluted by mundane and profane concerns. The sociology of scientific knowledge approaches science like Durkheim (1912/1954) approached religion; that is, with the intention of disenchanting the mysteries surrounding the sacred, and of revealing the social core of that which seems entirely pure and non-social.

This is where the analogy ends, though. The neo-Durkheimians have extended Durkheim's analysis into a general social theory of group cognitions. But so far, the sociology of scientific knowledge has not been able to develop a general social theory of science, although the vision of such a theory was the core of the original Strong Program. The only exception is structural constructivism that argues from a conflict and materialistic perspective (Collins and Restivo 1983a, 1983b; Collins 1989; Restivo 1988, 1989; Restivo and Collins 1982), but which is not widely recognized among mainstream constructivists. I believe there are several reasons for this unfortunate shortcoming. First, as I have repeatedly observed before, the field remains overconcerned with debunking orthodox realist epistemology. The typical structure of microsocial case studies of science is this: first describe the philosophical model of rational science, then proceed to show that philosophy cannot explain what is actually happening in science, and finally introduce social factors as an alternative to rational factors in accounting for the scientific process.[1] Social factors are frequently introduced only *after* it has been shown that rational factors miss actual scientific practice. The very term "social factors" implicitly retains the discredited epistemological dualism; as if there were "social" and "non-social" factors operating in science.

More recently, epistemological has been coupled with literary critique. The "reflexivist" group in the sociology of scientific knowledge (Ashmore 1989; Mulkay 1985; Woolgar 1988b) broadens the critique of philosophical realism to include a critique of empiricist textual practices. The attention turns to the rhetorical devices sociologists of scientific knowledge use to create the fiction that they are writing about the object of science. Reflexivism leads away from the study of science into self-referential textual narcissism. I would say that the important gains in the field have come from detailed empirical investigations conducted in the realist mode, not from epistemological critique or textual deconstruction.

Besides their preoccupation with epistemological and literary critique, there is a second reason for why social studies of science have failed to lead up to a general social theory of scientific knowledge. The vast majority of social studies of science are microstudies of individual laboratories, texts, controversies, and conversations between scientists. The style of these studies is descriptive and noncomparative. Implicitly, all science is treated as being the same, and there are usually no acknowledgements of possible temporal and disciplinary variations in scientific practices. A case in point is that many laboratory ethnographies have been conducted in biology labs. It is not just that microsocial case studies of particular laboratories or controversies might not be representative of science in general and thus lack external validity; although I do find the complete unawareness of such an elementary methodological problem somewhat bothersome. But a more serious problem, I believe, is that due to the absence of any comparative theoretical framework, case studies of science do not seem to add up to anything.

Studies of controversies, for example, simply reiterate time and time again that conflicts are not resolved by unambiguous experimental replications but by "contingent social factors." So far, a common result obtained in microsocial studies of science seems to be that realist epistemology is wrong because scientific knowledge is socially constructed and negotiated. Given the youth of the field and the long dominance of realism over the discourse on science, this result does represent a crucial epistemic gain. But even within the sociology of scientific knowledge, it is sometimes felt that it is time to move on, to abandon epistemological critique, and to revive the original thrust of the Strong Program; that is, to develop a comparative social theory of scientific knowledge.

The most important single reason for the present stagnation in social studies of science is this: *nothing is allowed to vary.*[2] Closure mechanisms, discursive practices, or the dynamics of knowledge construction and fact production are all treated as constants. In a way, science is treated as if it did not change over time, and as if scientific practices did not vary across fields and disciplines. But where nothing is allowed to vary, nothing can be explained and compared, for explanation and comparison presuppose covariation. Hence the narrative, nonexplanatory, and static mode of analysis predominant in social studies of science. Very characteristically, Karin Knorr-Cetina subtitles her laboratory ethnography, *The Manufacture of Knowledge* (1981), "An Essay on the Constructivist and Contextual *Nature* of Science" (my emphasis). Ironically, claiming to have discovered the social "nature" of science echoes the philosopher's claim to have discovered the "rational" nature of science—the only difference being social versus rational.

In order to develop a general and comparative social theory of science, I believe it is necessary to allow for *variations* in scientific practices. Constants must be converted into variables, and essentialist assumptions about the nature of science must be replaced by concepts that are sensitive to variation. But accounting for variations requires broadening the microperspective prevalent in social studies of science. I want to suggest that a general social theory of science must recover the mesoperspective on the organization of scientific communities that has been the conceptual core of the Mertonian paradigm. To be sure, this is not to say that we should also recover the Mertonian preoccupation with the institutional norms of science and with the role of reward and peer inspection systems in maintaining these norms. The critique of Mertonian normativism (e.g., Mulkay 1979) is well founded and justified. Most importantly, despite some declarations to the contrary (Gieryn 1982), the Mertonian framework is not very sensitive to the internal workings of science, to the actual "hows" of scientific production, and to the everyday contingencies of science-in-the-making. But I do believe that the microsocial critique of Mertonian normativism has gone too far in turning away from the mesocontexts of scientific production, such as patterns of community organization, systems of scientific communication, or levels of professionalization.[3] My basic argument is this: in order to develop a general theory of scientific production, we need to recover the mesoframework underlying the Mertonian paradigm, but we need not *uno actu* adopt its functionalist and normativist biases. What is needed is a general theory of scientific organizations that is sufficiently "macro" to account for disciplinary and temporal variations within and between scientific fields and, at the same time, is able to address the internal workings and contents of scientific production.

The theory of scientific organizations addresses the internal workings of science without losing sight of the structural contexts in which scientific work takes place. These larger contexts are not seen as separate from matters of content, as in Mertonian research, but are seen as determining how scientific work is conducted. The theory of scientific organizations is not an alternative to the case-study approach but offers instead a framework within which such studies can be interpreted and compared. The theory does not claim to explain the hypercomplexity of empirical reality, but is one way of reducing this complexity. The theory is thus guilty of simplification although, it is hoped, not of oversimplification.

One word about the models I will use throughout this chapter.

I hope the four-box format will not lead to the misunderstanding that this is a dull Parsonsian exercise of putting things into boxes. The models are not conceptual but explanatory, they are not meant as definitions or classifications but as explanations. Of course, these models are also simplifications and do not claim to cover reality in its entirety. They highlight and exaggerate differences in cognitive styles between various social and scientific groups. But instead of simply assigning cognitive styles to social groups, the models *explain* why it is that different social groups perceive reality in different ways and process information accordingly.

To illustrate the usefulness of an organizational approach to science, my goal in the remainder of this chapter is to show how some of the most central and significant topics and findings in social studies of science can be explained as outcomes of variations in the organizational structures of scientific fields. My purpose here is purely conceptual and heuristic. I do not claim to present a full-fledged organizational analysis of actual scientific fields. My aim is simply to demonstrate how the unsystematic and narrative microsocial evidence on science, when treated as variables, can be explained as covariates of social-structural and organizational variables. If we want to overcome the common practice of arguing from an isolated perspective without attempts at integration, it is necessary to show how our theory can account for the findings reported in microsocial studies of science as "special cases" of the theory. A theory can claim superiority only when it explains other approaches *and something more*. To this end, two conceptual cornerstones of the organizational paradigm must briefly be introduced: task uncertainty and mutual dependence.

"Mutual dependence" signifies the extent to which scientists are dependent on particular networks of collegiate control that organize the distribution of reputational and material rewards (Whitley 1984; Fuchs and Turner 1986). The more concentrated the organizational control over the distribution of reputational credits and over the allocation of material and symbolic resources for scientific production, the more dependent will scientists be on the particular professional establishments who administer this control. Conversely, if the level of concentration in the control over the resources for scientific production is low, then scientists may gain reputational credits and material rewards from a variety of organizational and extraorganizational sources, thereby making scientists less dependent on particular professional groups.

Presumably, high levels of mutual dependence between scientists generate rather firmly integrated and uniform ways of doing science. If scientists closely depend on each other for recognition and rewards, they are more likely to develop shared ways of approaching problems, of interpreting the evidence, and of defining the "important" issues in a given field. Conversely, low levels of mutual dependence are likely to support more loosely coupled, heterogeneous, and pluralistic research practices. For if mutual dependence is low, a variety of organizational, material, and symbolic resources for intellectual production—such as status positions, salaries, and technical equipment—are available, and hence, multiple ways of doing research may coexist.

"Task uncertainty" indicates the extent to which scientific production is routinized and predictable. Task uncertainty is low when problem definitions are fairly stable, when methods prescribe routine cognitive choices, and when there is a great deal of agreement on what constitutes legitimate and acceptable research. Conversely, the level of task uncertainty is high when problems and concepts are not clearly defined, when multiple methods for researching similar issues are available, and when the criteria for sound and significant contributions to knowledge are diffuse and controversial. Presumably, task uncertainty negatively covaries with mutual dependence. For when dependence is low, multiple ways of doing research may coexist, and consensus over the correct ways of doing science will be fragmented.

I shall use this rudimentary organizational model of scientific production to address three crucial topics from my previous discussion of microsocial studies of science: the "idiosyncratic" nature of scientific production, the social dynamics of fact production, and scientific controversies. Again, my purely heuristic goal here is to illustrate how an organizational model of scientific production might serve as a more powerful and comprehensive framework for explaining what microsocial studies of science only describe as the "essential nature" of scientific reasoning. I shall also deal with the relationship between mundane and scientific knowledge, and suggest that the same theory can explain differences in cognitive styles—for mundane as well as scientific groups.

THE IDIOSYNCRATIC NATURE OF SCIENTIFIC PRODUCTION

A popular and recurrent theme in microsocial studies of science, particularly in ethnographies of laboratory life, is the "situationally contingent" and "indexical" character of scientific work (Knorr-Cetina 1981, 1983; Garfinkel, Lynch, and Livingston 1981; Latour 1980; Lynch 1985). Borrowing from ethnomethodological analyses of interactions-in-contexts, many lab ethnographers assume scientific work to depend on the locally available resources and idiosyncratic research cultures predominant at particular laboratories. Whereas epistemology normatively assumed that scientific work was governed by general and agreed-upon rules that standardized research activities and made work outcomes compatible across scientific communities, ethnographic studies of laboratory life stress the occasioned, opportunistic, and circumstantial nature of scientific work. Laboratories not only differ in terms of the technical equipment and material resources available to scientists, but they also establish unique cultures and traditions of research. Lab ethnographers believe that the local circumstances of scientific production are not accidental to the outcomes of research; rather, these outcomes are held to be thoroughly constituted by the local customs of scientific production. In this view, the selective decisions structuring the research process are profoundly shaped by the particular instruments used in a lab, by locally specific measurement techniques, and by the unique materials and substances preferred at particular laboratories:

> This contextual location reveals that the products of scientific research are fabricated and negotiated by particular agents at a particular time and place; that these products are carried by the particular interests of these agents, and by local rather than universally valid interpretations; and that the scientific actors play on the very limits of the situational location of their action. In short, the contingency and contextuality of scientific action demonstrates that the products of science are hybrids which bear the mark of the very indexical logic which characterizes their production. (Knorr-Cetina 1983:124)

Several critics have correctly objected that ethnographers of laboratory life greatly exaggerate the locally occasioned and idiosyncratic character of scientific work. Gad Freudenthal (1984:289) and Harriet Zuckerman (1988:546) note that the radical localist hypothesis has difficulties explaining why multiple discoveries and anticipations occur so frequently in science, and Tom Gieryn (1982:290f.) diagnoses the inability of the localist model to account for the importance of translocal standards for evaluating scientific work. Indeed, if research practices and outcomes varied significantly from location to location, anticipations and multiple discoveries would occur by chance alone, which contradicts the observed high frequency of parallel findings (Freudenthal 1984; Hagstrom 1965). Similarly, if research results were entirely contingent and idiosyncratic, there would be no comparative standards for evaluating and rewarding scientific work. More importantly, scientists could not be induced to incorporate other scientists' findings into their own work. Localistic research cultures would merely coexist without attempts at using other scientists' work as resources for one's own work. Hence, the localist hypothesis is incompatible with the very core dynamics of fact production in science.[4]

The localist hypothesis greatly exaggerates the idiosyncratic character of scientific work and underrates the importance of translocal and comparative standards for assessing research practices and outcomes. This overemphasis on unique local laboratory cultures reflects the absence of any adequate notion of scientific community organization in microsocial studies of science. The fact that scientific knowledge is socially constructed need *not* imply that science is entirely idiosyncratic and dependent on local contexts (Galison 1987:254), for cars are socially produced as well, but the car production process is rather standardized and does not vary a great deal from plant to plant. I believe the social production metaphor of scientific knowledge is very useful, but since Marx, one can know that production processes do vary. To be sure, this is not to say that epistemology was right in insisting that science was a rational process because it followed universal standards of methodological propriety. But ironically, orthodox epistemology and lab ethnography commit the same conceptual fallacy: they both claim to describe the "nature" of science, and they both do not allow for variations in scientific practices.

Such pointless essentialist debates over the nature of science can be avoided if we treat scientific practices as variables instead of as constants. There is no reason to assume that all science is the same and that research practices do not vary across disciplinary fields and over time. That is, scientific fields[5] differ in the extent to which research practices are standardized, formalized, and predictable; just as organizations employing various technologies differ in their degrees of structural formalization (Thompson 1967; Perrow 1967). Instead of postulating that all sciences are always rational or always idiosyncratic, allowing for variations in scientific practices implies that some fields, at certain times, look more like the standard model of rational science, whereas other fields institutionalize more localistic and idiosyncratic research patterns.

In terms of the organizational variables introduced above, task uncertainty and mutual dependence, we would expect a high degree of standardization and formalization in scientific practices under conditions of low uncertainty and high dependence. Highly standardized research practices will emerge when scientists very closely depend on other scientists for recognition and rewards, and when there is a great deal of agreement over the proper ways of doing research. In this situation, there is less room for idiosyncratic practices and local research cultures. High mutual dependence means that control over scientific production will be rather concentrated and centralized so that deviant and idiosyncratic ways of doing research cannot be sustained by fragmented resource pools. Under these conditions, there will not be a great deal of controversy and disagreement over the correct ways of doing science so that task uncertainty will be rather low. That is, the rational realist model of science with its normative emphasis on cognitive standardization and routinized methodologies does not describe the nature of science *per se*, but scientific fields with high degrees of material and symbolic integration. In this view, it is no accident that realism has traditionally regarded physics as the exemplar of a "mature" science, for physics is a rather closely integrated field with more standardized research practices (Whitley 1984).

Corroborating this point empirically, Collins and Restivo (Collins and Restivo 1983b; Restivo and Collins 1982) document for the cases of Greek and modern mathematics that increasing density and dependence in the mathematical community led to a stronger emphasis on formal symbolisms and logical rigor. While a more individualistic mode of mathematical production leads to interpersonal rivalries and idiosyncratic practices, higher density in the community of mathematicians favors abstract proofs and internal systematization. "Pure" mathematics emerges under conditions of increasing professionalization and historical continuity (Restivo 1990). The more collective orientation of the mathematical field does not eliminate competition and conflict, but restricts their scope within a body of shared cognitions. Collins (1989:132ff.) finds the same pattern in the history of philosophy: under conditions of increasing social density among networks of philosophers, competition between various camps leads to higher levels of abstraction, to attempts at formal proofs, canons of logic, and formal epistemological rationalization.

Conversely, more idiosyncratic and locally unique research practices are more likely to occur in fields with comparatively low mutual dependence and high task uncertainty. Low dependence allows for the coexistence of multiple approaches and paradigms, for the control over scientific resources is rather decentralized and dispersed. Under these conditions, various and conflicting ways of doing research can be materially secured so that the overall level of agreement will be rather low and task uncertainty high. In the absence of widely shared and standard ways of doing research, research decisions will be more dependent on the contingent circumstances of scientific production. The discretion of individual scientists and the influence of local groups over scientific production will be rather high, while the control of translocal communities will be weak. That is, low dependence and high uncertainty are conducive to the locally idiosyncratic and occasioned research practices that microsocial studies of science regard as constitutive of the very nature of science. But neither realism nor microsocial studies describe the nature of science *per se*. Rather, they describe the opposite poles of what should be treated as a *continuum* of cognitive standardization and routinization. Figure 4.1 arranges the extreme poles of that twofold continuum.

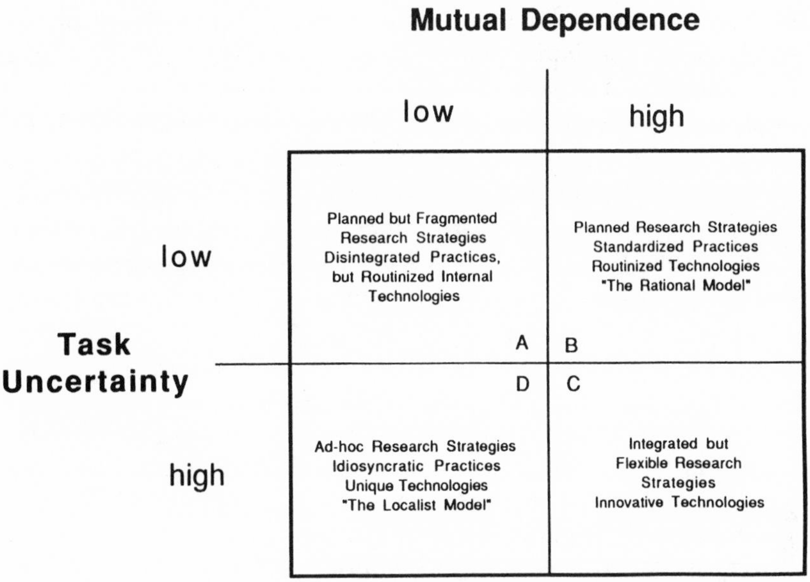

Mutual Dependence

FIGURE 4.1   A Model of Research Practices

It appears that these relationships between structure and cognition are confirmed by research on general organizations. The low dependence/high uncertainty condition is present in loosely coupled organizations with ambiguous goals and conflicting definitions of reality. This type of "organized anarchy" has been described by "garbage can" and institutionalist approaches to organizations (March and Olsen 1979; Meyer and Rowan 1977). It appears that higher-level educational institutions, for example, fit this pattern rather closely. In such organizations, practices are more informal, ad hoc, and idiosyncratic; they are not governed by a scientific method of rational administration (which would be the equivalent to a "methodology" in science). In such organizations with low dependence and high uncertainty, the formal system does not really determine what is going on. The actual policies are due more to informal negotiations and flexible adjustments.

Conversely, low uncertainty organizations are those which look more like classical Weberian bureaucracies. Examples are the DMV, most prisons, or welfare agencies. In such organizations, there is a great deal of paperwork and a lot of formal rules and regulations (or, in an analogy to science, a "rational methodology" of administration). Weberian bureaucracies do their work in very similar ways, for there is not much local variability and idiosyncratic politics. This is especially so in the pooled production-type organizations (Thompson 1967), where a central agency allocates resources to independent branches, as in post offices.

In any case, I believe we should treat scientific (and organizational) practices as variables rather than constants. A powerful theory explains why and under what conditions this rather than that is the case. This is only possible if we allow for variations.

THE SOCIAL DYNAMICS OF FACT PRODUCTION

According to discourse analysis and reflexivism, the production of facts is largely a matter of rhetoric and literary conversion (Gusfield 1976; Myers 1985; Woolgar 1980, Ashmore 1989). Following a currently fashionable theme in poststructuralist semiotics, writing is idealistically seen as the process of constructing cultural signs and artifacts through literary inscription. While scientists in the laboratory are mundane reasoners fabricating knowledge claims in a largely idiosyncratic, indexical, and ad hoc fashion, the standard research article presents objective findings that have erased all traces of social construction, correspond to reality, and were obtained by following rational procedures. That is, discourse analysis believes that facts are constructed by converting the indexical logic of laboratory reasoning into the decontextualized style of the conventional research report. Facts are held to be first and foremost *literary* and rhetorical products; they are not assumed to have any reality apart from their inscription in the research article.

Following Latour (1987), however, I have analyzed fact production as an eminently social, not just literary, process. Facts emerge when scientists use statements as black boxes or premises for their own work. To induce other scientists to gradually transform statements into facts, researchers have a variety of supporting textual and nontextual resources at their disposal, such as networks of quotations, numbers, graphs, laboratory technologies, computers, and reputational capital. These supporting agents increase the costs for objecting to particular claims and, in this way, increase the probability of a statement being transformed into a fact. The stronger the textual and nontextual agents supporting a statement, the more likely are other scientists to use that statement as a black box on which to build their own statements.

Although Latour's (1987) model does not explicitly allow for variations between fields, his constants can easily be converted into variables. Most importantly, Latour argues that facts emerge from social support networks. But we know from network analysis that such networks differ in their density and power structures (see Wellman and Berkowitz 1988). There is no *a priori* reason to assume that all scientific fields have equally strong support networks and resource pools to back up statements. It seems reasonable to assume that high-energy physics differs on these variables from history, and that experimental areas differ from discursive and textual fields. Consequently, we would expect their cognitive styles to differ, so that some strong and powerful fields produce solid black boxes whereas others engage in informal and pluralistic intellectual activities.

Scientific fields may differ in the strength and level of concentration of the symbolic and material resources mobilized to support statements (Whitley 1984; Fuchs and Turner 1986). Generally, the stronger and more concentrated the symbolic and material resources in a given field, the higher the level of mutual dependence between scientists. In fields with comparatively low levels of resource concentration, such as sociology, the degree of mutual dependence will be rather low, since multiple scientific establishments, "schools," or paradigmatic communities will coexist.[6] But if resource concentration is rather high, such as in the more "mature" sciences, scientists depend more closely on the particular scholarly networks that control these resources. Traweek (1988) has described this situation for the case of high-energy physics. *Fields with high levels of resource concentration and mutual dependence are more likely to produce facts than fields with low degrees of concentration and dependence.* Recall that fact production is a social process: only other scientists can transform statements into facts. And consequently, the stronger and the more concentrated the resources are that scientists may use to support their claims—the higher the level of mutual dependence—the more likely are facts to emerge. If scientists closely depend on each other, they are more likely to utilize each others' statements as resources for their own work. Under these conditions, there is a

strong feeling of group solidarity and a firm belief in the superiority of its worldview. Collective controls over practitioners are
rather tight, and there is not much tolerance for deviance and
dissent. Fields that produce facts are perceived as rather mature,
enjoy a great deal of social prestige, and will be conceptually and
methodologically integrated. There will be a strong "scientistic"
confidence in rational and neutral ways to produce true knowledge
about the external world. Fact-producing fields will not be very
concerned with the metaphysical "presuppositions" that structure
their worldviews. There is no elaborate "philosophy" that deals
with foundational issues or normative problems. Rather, such
fields will display a strong pragmatic future orientation toward the
gradual approximation of knowledge to reality. Consequently,
such fields will not greatly emphasize their historical traditions
and origins (Kuhn 1970). History is strictly separated from
systematics.

Conversely, if the material, organizational, and symbolic resources for research are dispersed and distributed more widely to
multiple paradigmatic networks within a given field, the social
pressures inducing scientists to turn each others' statements into
facts are much weaker. In these fields solid black boxes are less
likely to be closed, for the supporting agents and resources which
back up statements are less concentrated. There are fewer and
weaker inducements to motivate other scientists to accept one's
statements as the premises for subsequent research, for researchers
belong to distinct groups, schools, and traditions that might not
even take notice of each others' work. *Such fields are more likely to
produce "conversation" and hermeneutics than facts and science.*[7]
If low levels of resource concentration and mutual dependence
allow for the coexistence of multiple paradigmatic communities,
the exchanges between these communities will be of a more discursive, mutually critical, and informal sort. The various independent
groups and subspecialties lack the resources to induce other groups
to use their work as premises for further research. As a result,
knowledge will be seen more as an active social construction than a
neutral representation of objective reality. Since they are only
loosely coupled, organizationally and materially fragmented fields
do not exert strong social pressures on practitioners, and so the

incentives to incorporate other scientists' research in one's own work are fairly weak.

As a result, the cognitive style of conversational fields will be more interpretive, more skeptical of firm and absolute foundations for knowledge, and more inclined to find its identity in constantly reevaluated historical traditions rather than in solid facts and truly scientific methods. The history of such fields merges with their systematic research efforts. Ongoing philosophical debates on the foundations and transempirical implications of conversational fields create a strong nonscientistic and discursive intellectual habitus. An example is the continuing debate on the "classics" in conversational and multiparadigmatic fields (Alexander 1987). In fragmented and loosely coupled fields, constant reinterpretations of classical scriptures provide a "paradigm substitute" that defines at least a minimal consensus on the basics of the field, and creates a however weak sense of belonging to the same group. Due to the coexistence of multiple paradigmatic communities, knowledge will be seen more as a selective construction than as a neutral mirror of reality. Each separate community celebrates its own intellectual heroes and traditions. Research is seen to be based partly on contingent decisions and not only on the imperatives of reality and rationality proper. Methods are more likely to be interpreted as "soft" pragmatic tools that not only help to discover but also thoroughly construct whichever reality is being examined. Weak collective and translocal controls allow for high practitioner autonomy and independence. These fields establish "idiosyncratic" and "locally occasioned" research practices that are often not compatible with each other. The role of active agency and creative imagination in the fabrication of knowledge is stressed. Theoretical rationales will be seen as "approaches" and "perspectives" among other possible approaches and perspectives, not as the only adequate ways to scientifically explain reality.

If we combine the now familiar organizational variables of mutual dependence and task uncertainty, we again obtain a model of scientific fields producing different types of knowledge and cognitive interaction. Note again that this highly simplified and schematic model is not meant to cover actual scientific fields, but represents an attempt to explain variations in cognitive styles by

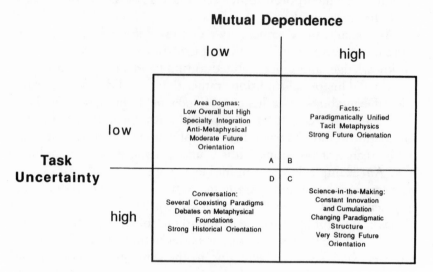

FIGURE 4.2   A Model of Cognitive Styles

corresponding variations in the organizational structures of scientific fields. In addition to the two types of cognitive styles already discussed, facts and conversation, figure 4.2 also expects "science-in-the-making" and "area dogmas." Science-in-the-making refers to the elitist and exclusive networks of scientists working in the innovative core areas of a discipline. Although these networks are typically closely coupled and recruit their members mainly through master-apprentice ties (Zuckerman 1977; R. Collins 1989), the comparatively high level of task uncertainty in areas of novel knowledge means that epistemic standards and collective controls will be less rigid and formal. Due to small group size, exclusive social status, and high mutual dependence there is a strong sense of solidarity. At the same time, collective inspection and controls are tempered by the necessity to grant considerable autonomy and discretion to scientists working on highly uncertain and innovative problems. Innovations cannot be produced by following strict methodological canons, and so scientists working on innovations must be given considerable work discretion and autonomy.

Area dogmas indicate subspecialties in largely disintegrated disciplines, such as demography or criminology in sociology.

These subspecialties have their own reputational and organizational systems and are not very dependent on and concerned about the overall field. Under these conditions, the integration of the larger discipline is not very high, but intragroup exchanges will be much more frequent and dense. Such groups work on fairly specific problems and narrow questions, and "progress" is made as piecemeal improvements in knowledge and methods. Participants are cautiously optimistic about advances in their own fields, but are rather skeptical of and detached from the larger discipline.

MUNDANE AND SCIENTIFIC KNOWLEDGE

So far, I have not addressed the issue of whether the organizational approach to scientific knowledge can also explain variations in the cognitive styles of mundane social groups. The sociology of scientific knowledge has made no attempts at applying its framework to mundane knowledge, although its main point—that nothing special is happening in science—makes sense only if we have a theory that explains differences and similarities between science and ordinary discourse. By confining itself to narrative descriptions of individual settings and events in science, the sociology of scientific knowledge implicitly reinforces the special nature science has been given by realist epistemology. To overcome the isolation of the field, it is necessary to show how the organizational approach to science can be integrated with a sociology of mundane knowledge. Science can then be seen as a special case of knowledge production, which helps disenchant the ideological mysteries surrounding modern science and protecting it from critical public inspection. Neo-Durkheimian grid-group theory offers an excellent way to accomplish this goal. It turns out that the organizational cum neo-Durkheimian predictions about epistemic styles do not differ much for science and ordinary reasoning.

The general argument underlying the model summarized in figure 4.2 draws upon Durkheim's (1893/1965) sociology of groups and group cognitions. Durkheim argued that cognitive styles covary with the degrees of social density or cohesion found in groups and entire societies. Macrostructurally, the famous distinction between "mechanical" and "organic" solidarity was meant to explain the changes that worldviews and value systems undergo when societies grow more complex and differentiated. Small tribal societies with kinship bonds and segmentary differentiation ("mechanical solidarity") generate totemistic religions with highly reified and particularistic symbolisms, whereas complex modern societies with functional differentiation ("organic solidarity") create more abstract and universalistic symbolisms.

Microstructurally, Durkheim (1912/1954) explained the ritual production of cognitive order and moral solidarity. The social rituals performed at periodically staged group gatherings create and reinforce emotional ties between members, and charge collective symbols with emotional energy to sustain a sense of belonging and collective identity.

The neo-Durkheimians have systematized Durkheim's general argument into what has come to be known as a grid-group approach to social structure; with "grid" representing the vertical lines of social stratification and "group" representing the horizontal levels of social integration. Originally developed by Mary Douglas (1966, 1970) in her social anthropology of tribal societies and their moral cosmologies, grid-group analysis has—in some version or other—been applied to such various fields as the ritual theory of class cultures (Collins 1988), the sociology of art and music (Bergesen 1988), sociolinguistics (Bernstein 1974–77), ethnic relations (Fuchs and Case 1989, 1990), and the sociology of language games or lifeforms (Bloor 1983). Grid-group analysis intends to explain variations in cognitive styles, moral views, and social perceptions by corresponding variations in the degrees of social solidarity or density among groups; and hence it is directly relevant for any attempt to develop a general social theory of (scientific) knowledge. I shall concentrate on the horizontal or "group" aspect of grid-group analysis, for the organizational variable of mutual dependence covers much the same dimensions of social structure as Douglas's group variable.

"Mutual dependence" corresponds to what Douglas (1966, 1970) calls the "group" aspect of social structure, and also to what R. Collins (1975, 1988) calls the level of "social or ritual density" found in groups and classes. Under conditions of high mutual dependence or social density, groups are rather homogeneous and well integrated, develop a strong sense of group membership and collective belonging, and establish sharp boundaries between members and nonmembers. The more dependent group members are upon each other, and the more closely and frequently they interact with members rather than with outsiders, the stronger will be the shared sense of belonging, and the tighter will be the emotional and cognitive ties between members.

Douglas argues that such groups are very pollution-conscious, develop a strong sense of internal ritual cleanliness and purity, and cautiously seek to maintain very rigid demarcations between members and outsiders. Under conditions of high mutual dependence or high social density, groups are rather intolerant toward internal deviance, emphasize conformity and consensus rather than competition and conflict, and place the welfare of the group as a whole over the interests of individuals. Hence, densely structured groups closely supervise and control members' activities, which is conducive to a rather authoritarian and stratified internal organization, or high "grid."

Due to the high level of shared commitments and beliefs, communication between members is rather restricted in the sense that it occurs mainly through rigidly interpreted symbols and stereotypical formulas (Bernstein 1974–77). Since the level of solidarity or cohesion is high, collective symbols directly express group membership and, therefore, are worshipped as sacred objects. Usually, there are sharp distinctions between profane and sacred matters, with profane trespasses into the sacred being subject to severe penalties. In densely integrated groups, members are ascriptively identified by the ways they talk, dress, and interact with each other, and these sacred ways of expressing the group identity must not be polluted by profane, "foreign," and "inferior" influences coming from the outside.

The thrust of neo-Durkheimian theory is to explain intergroup variations in cognitive styles and cultural outlooks by underlying variations in the degrees of mutual dependence, social density, or "group." Groups with high levels of mutual dependence or social density generate rather uniform, dogmatic, and inflexible cognitions and moral beliefs. The sacred symbols representing group membership are taken at face value; they are interpreted as reified signs of collective identity, not as abstract and conventional symbols (Collins 1975, 1988). Collective beliefs are given standard interpretations without much tolerance for dissident views. The worldviews of these groups are organized very coherently, and there is not much room for exceptions and inconsistencies. The logical modality of such belief systems is deterministic necessity, not probability or contingency (Luhmann 1984). Group interpretations must be kept pure and stable, while changes and innovations are looked at with suspicion. The group closely watches over the "right ways" of thinking and acting, and there is not much concern for individual creativity and spontaneity.

Hence, the cognitive style of dense social groups is rather formalistic (R. Collins 1975:166). Thought must meet strict standards of appropriateness, and form is considered more important than content. Tradition is maintained as a value *per se*. In Bloor's (1983:141f.) terms, socially dense groups are "monster-barring" collectivities who hate anomalies and strangers. Since high social density implies strict demarcations of insiders from outsiders, collective cognitions employ simple binary distinctions between inside and outside, good and bad, sacred and profane, superior and inferior. That is, the worldviews of densely organized groups are in a fundamental sense "prejudiced" (Fuchs and Case 1989, 1990). Since the group is more important than its individual members, cognitions typically refer to impersonal and supraindividual, often religious forces, whereas individual interests and agency are considered less important.

From a neo-Durkheimian perspective, groups that are structured in similar ways produce similar kinds of cognitive styles; be they groups of scientists or mundane actors. That is, *groups employ a "fact mode" of reasoning whenever they are cohesive and routinized, while a "conversation mode" is typical of loosely coupled groups with a great deal of individualism and independence.* A fact mode of reasoning has great confidence in its superior rationality and objectivity. Knowledge is seen dogmatically as expressing the way things really are. Cognitive practices are structured rather formally and rigidly. There is one right way of doing things, and it must be kept pure and immune against heresy. A conversation mode of reasoning, on the other hand, is less rigid and dogmatic. Knowledge is seen as selective and contingent. There is more skepticism than certainty, more tolerance than dogmatism, and the contingency of interpretations is stressed. Knowledge is more construction than representation, and many constructions are seen as possible. A conversational mode of reasoning is more reflexive, relativistic, and informal, while a factual mode is more reified, restricted, and methodological. The mundane correlate to scientific facts is the symbolic reification of sacred symbols, and the "conversational" style of disintegrated scientific fields corresponds to the intellectual cosmopolitanism and pluralism of mundane groups with weak collective and emotional ties (Fuchs and Case 1990).

Let us now reconsider figure 4.2. Although this model of cognitive styles is not meant to cover actual scientific fields, it is probably not incorrect to assume that box B roughly covers the organizational structures of the "hard" and "mature" natural sciences, whereas box D covers the "soft" and multiparadigmatic social sciences and humanities.[8] Generally, the natural sciences have institutionalized higher levels of mutual dependence, and hence are cognitively more unified than the more loosely organized social sciences (Whitley 1984; Hargens 1988; Levitt and Nass 1989; Lodahl and Gordon 1972). The most important single reason for this is probably that most social sciences, and all the humanities, are nonexperimental and hence do not use very expensive equipment or establish much copresence between scientists at the workplace. Due to the high costs of lab equipment, experimental methods tend to concentrate the material means of scientific production, and they also draw various scientists together in one place. As a result, density or mutual dependence increases, and extended periods of copresence create similar and more uniform ways of perceiving the world (Fuchs 1989).

If this argument holds, we would also expect significant differences between the traditional philosophy of the natural sciences, realism, and the philosophy of the social sciences.[9] More specifically, we would expect realism to structurally resemble the worldviews of socially dense mundane groups with high levels of mutual dependence, whereas the philosophy of the social sciences should resemble the more loosely structured worldviews of groups with low social density. These differences are summarized in figure 4.3.

Realism has traditionally regarded physics as the model for a "mature" science. According to realist epistemology, mature sciences look a lot like the worldviews of groups with high social density (see Hooker 1987:62–84). Science is a very coherent sys-

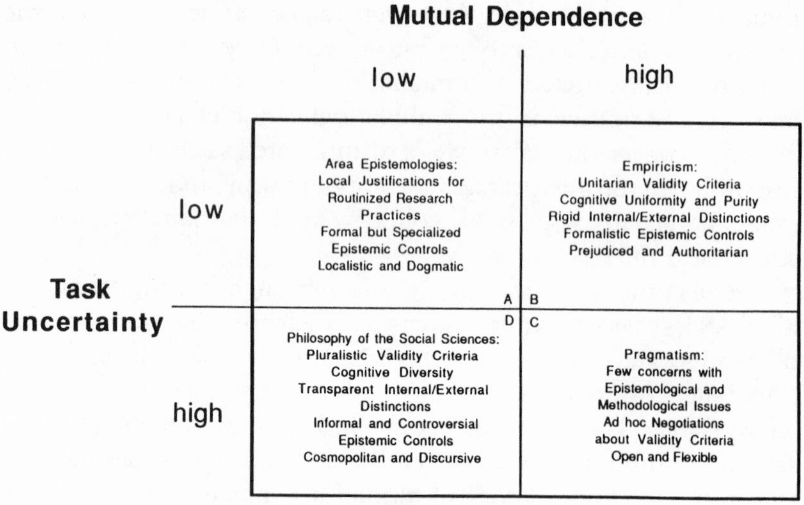

Mutual Dependence

	low	high
low	Area Epistemologies: Local Justifications for Routinized Research Practices Formal but Specialized Epistemic Controls Localistic and Dogmatic	Empiricism: Unitarian Validity Criteria Cognitive Uniformity and Purity Rigid Internal/External Distinctions Formalistic Epistemic Controls Prejudiced and Authoritarian
	A | B	
	D | C	
high	Philosophy of the Social Sciences: Pluralistic Validity Criteria Cognitive Diversity Transparent Internal/External Distinctions Informal and Controversial Epistemic Controls Cosmopolitan and Discursive	Pragmatism: Few concerns with Epistemological and Methodological Issues Ad hoc Negotiations about Validity Criteria Open and Flexible

(Task Uncertainty, with "low" and "high" on the left axis)

FIGURE 4.3   A Model of Epistemologies

tem of deductively interrelated theoretical propositions with rigid rules for operationalization and testing. Like the worldviews of densely organized social groups, orthodox realism establishes fairly absolutistic and authoritarian standards for valid knowledge. Correct cognition must always follow the laws of logic and the operational rules for translating theoretical into observational terms. The rules and laws of correct scientific thought are held to be universal; they serve as the normative model for all cultural fields that claim to produce objective knowledge. Hence, knowledge that cannot ultimately be reduced to elementary protocol sentences is not knowledge at all but metaphysics or fiction. That is, like the rigid worldviews of socially dense groups, the epistemic ideal of realism is *purity*: pure logic and undistorted perception are worshipped as the sacred foundations of knowledge.

In a certain sense, then, realism is also a very "prejudiced" belief system. The knowledge expressed in the arts, in music, literature, or ethics does not really deserve the title of "knowledge." Science is seen as the ultimate paradigm for all rational discourse and as structurally different from and superior to mundane ways of knowing. Like the worldviews of socially dense groups, realism draws sharp distinctions between science and fiction, knowledge and illusion, inside and outside. Most importantly, realism strictly separates "internal" from "external" aspects of science. Like the worldviews of dense social groups, realism is very boundary-conscious and abhors transgressions from the outside. The internal side of science consists of the pure and sacred forces of Reason and Reality, whereas the external side of science consists of the messy and profane forces of social influences and ordinary interests. The inside of science must not be polluted by the intrusion of external factors into the research process. Also, the pure force of Reason is impersonal and supraindividual, just like the quasi- religious and transcendental forces worshipped by densely structured social groups.

In short, there are significant structural resemblances between the worldviews of socially dense groups and the realist philosophy of the natural sciences. Following the general lines of neo-Durkheimian theory, I suggest that these structural resemblances are due to underlying similarities in the social structures of densely organized "ordinary" groups and closely integrated scientific communities or research specialties. Generally, the higher the levels of mutual dependence or social density, the more reified, standardized, pollution-conscious, and formalistic the collectively shared cognitive styles. If this general argument is correct, we should observe similar structural correspondences between the organizational structures of the social sciences and the cognitive structures of their philosophy.

Consider again figure 4.2, box D. I would say that very roughly and schematically, low mutual dependence and high task uncertainty characterize the organizational structures of the social sciences and humanities; although undoubtedly, there is also considerable variation *within* these fields as well. In sociology, for example, there are a number of diverse paradigmatic communities controlling the resources for scientific production. Therefore, the overall level of mutual dependence is rather low. Consequently, we would expect "the" philosophy of the social sciences to be structurally rather different from realism. Of course, there is not just one philosophy of the social sciences, but this fact only corroborates the neo-Durkheimian argument: comparatively loose social and organizational structures generate and sustain rather diversified cognitive orientations.

Despite several largely unsuccessful attempts at introducing realism as *the* philosophy of the social sciences, the field is by no means symbolically unified. At present, various metatheories contest each others' views of the "foundations" of the social sciences, such as hermeneutics, positivism, macrostructuralism, and micro-approaches to human agency. As opposed to the universal and unitarian cognitive schema realism has developed for the natural sciences, sociology, for example, has come to be viewed as an essentially "multiparadigmatic" field based on various "presup-positional" decisions (Ritzer 1980; Alexander 1982–83). Instead of conceptual unification, paradigmatic diversity characterizes the philosophy of the social sciences. Sociology is seen by many as a discursive and literary rather than as a deductive and scientific enterprise. Comparatively loose, vague, and highly controversial standards for producing and validating knowledge are advocated by various camps. Especially since the interpretive turn sociology underwent in the mid-seventies, concepts and methods have come to be regarded as constructive devices and pragmatic tools rather than as neutral means to arrive at accurate representations of reality. That is, the philosophy of the social sciences has a less reified and formalistic understanding of conceptual and methodological instruments. The role of active agency in producing knowledge is stressed, and there is great variety in "approaches" and "perspectives" instead of a unified paradigm.

As opposed to realism, the philosophy of the social sciences draws no sharp distinctions between "internal" and "external" aspects of scientific knowledge. Ever since Max Weber's famous postulate of value neutrality, there has been a continuing debate on the impact of interests, values, and personal commitments on scientific production. While scholars on the more positivistic side of the epistemological continuum advocate a strict separation between facts and values, science and politics, or truth and morality, critics argue that since sociologists study social worlds of which they are members, such clear-cut separations are impossible or even undesirable. Critical Theory, for example, advocates an explicit commitment of the researcher to the norms of emancipation and utopian communicative democracy. Ever since sociology emerged as a discipline, some scholars have regarded it as a practical means for social reform, while others have envisioned a "pure" science of society modeled after the natural sciences.

Paradigmatic diversity, then, characterizes the philosophy of the social sciences. The ongoing and probably irresolvable debates on such fundamental issues as what constitutes social worlds, can sociology be a science, what methods are appropriate for the study of society, is the micro more or less fundamental than the macro, is agency more important than structure, etc., have created a sense of "anything goes" in sociology. No paradigm is universally accepted, and so the philosophy of the social sciences has gradually moved toward a tolerant and pluralistic relativism which fits rather well into the broader historical traditions of intellectual cosmopolitanism and individualistic liberalism prevalent in the field (Turner and Turner 1990).

From a neo-Durkheimian perspective, the comparatively pluralistic, individualistic, and relativistic self-understanding of the social sciences corresponds to their rather loose and fragmented organizational structures. Since in the social sciences, the level of mutual dependence—if broadly compared with that of the natural sciences—is rather low, cognitive practices are not highly standardized and formalistic. Hence, diversity, individualism, and cosmopolitan openness are valued more than rigid scientistic controls over the "right ways" of doing research. Generally, low levels of mutual dependence or social density generate rather cosmopolitan and individualistic worldviews stressing tolerance, diversity, and creativity; whereas high levels of dependence or density are conducive to more reified and formalistic worldviews stressing uniformity, purity, and orthodoxy (R. Collins 1975). Facts and rigid epistemic controls emerge when scientists closely depend on each other for resources, rewards, and recognition. Conversely, conversation, relativism, and intellectual pluralism originate in loosely coupled disciplines with multiple opportunities for scientific production.

From this strictly sociological perspective, the mysterious philosophical problems of relativism and reflexivity that still haunt sociologists of scientific knowledge appear as pseudoproblems. Relativism, or the absence of secure epistemic foundations, is the ideology of highly fragmented and individualistic social groups with low internal solidarity and cohesion. When mutual dependence is low and consensus at best partial, then multiple perspectives compete with each other, and each develops rationales and finds people to convince. *"Anything goes" does not describe the absence of foundations, but the lack of social solidarity.* Relativism is not a problem for closely coupled communities since these establish stronger pressures on conformity and have less tolerance and fewer resources to sustain prolonged deviance and dissent. Solving the relativism problem is not at all a philosophical, but a sociological, task.

The same holds true for reflexivity. In the sociology of scientific knowledge, reflexivity is seen to arise from the nature of its subject matter, that is, the construction of knowledge claims (Ashmore 1989). This is an oddly realist argument, given that realism is the prime target of reflexive deconstruction. Reflexivism wants to replace the realist implications of conventional representational practice by multivocal discourse that displays the secrets of its own construction while telling its story. Reflexivism is thus a philosophical (antirealist) position and a deconstructive textual practice.

From a sociological perspective, however, reflexive self-inspections are more likely to occur in loosely coupled conversational fields than in more integrated fact-producing fields. Conversational fields view knowledge more as a social construct than as a neutral representation, and so they stress the role of active agency in fabricating claims. Since a variety of approaches and perspectives exist in such fields, the selectivity of any one perspective can more readily be seen. In densely structured factual fields, however, reflexivity is less of a problem, for there is more confidence in scientific rigor and methodological precision. Such fields have to be *told* that their knowledge is a contingent social construct, for their *conscience collective* is more realist and scientistic.

Of course, in its present version this argument is rather schematic and greatly simplifies the actual organizational and intellectual diversity in the sciences. But I do believe that unlike microsocial studies of science, the argument proposed here does provide a strong explanatory and comparative framework for a general *social* theory of (scientific) knowledge. Let me illustrate this crucial point with one final example: scientific controversies.

CONTROVERSIES AS NORMAL ACCIDENTS

As we have seen in the previous chapter, the study of controversial science assumes a prominent position in the sociology of scientific knowledge. The Empirical Program of Relativism has selected controversies as its special domain because controversies are held to reveal the social and political "nature" of science most clearly. The main outcome of case studies of controversial science is that controversies are not and cannot be resolved by the impartial forces of Reason and Reality because during controversies, the voices of Reason and Reality are many. Hence, it is generally concluded that "social factors" operate as closure mechanisms (Collins 1985).

But despite their strategic importance in social studies of science, no systematic attempts have as yet been made at explaining why controversies occur to begin with. Since nothing is allowed to vary, it is implicitly assumed that episodes of controversial science are equally likely to occur in all scientific fields, that the social processes of conflict resolution do not vary across disciplines, and that all controversies are closed through much the same social factors, independent of time and patterns of community organization. That is, just as ethnographers of laboratory life postulate an indexical and idiosyncratic *nature* of science, so are studies of contemporary scientific controversies largely indifferent toward possible historical and disciplinary variations in scientific practices.

But again, there is no good *a priori* reason to assume that the causes and resolution processes of controversies are constants. Nor is there any reason or evidence to believe that the social factors responsible for closure are the same throughout all of science. Kuhn's (1970) account of scientific change is not very helpful, either, for he believes that change is ultimately driven by the sheer accumulation of empirical anomalies. But the ways in which groups and communities respond to "anomalies" is not given and fixed, but critically depends on internal social organization (Bloor 1983:142). Kuhn is too empiricistic and not sociological enough when it comes to explaining paradigm shifts, for cognitive inconsistencies still play a dominant role in his model (Restivo 1983).

With Collins and Restivo (Collins and Restivo 1983b; Restivo and Collins 1982; Collins 1989), I hold that *competition* is the motor behind scientific change. Since innovative contributions are the ones that are recognized and rewarded the most, scientists compete over claims to novel knowledge. But the crucial point is that various *kinds* of change are possible, and that the type of change triggered by competition depends on the structure of the relevant group. From an organizational perspective, conflicts and controversies are structurally more likely to occur in certain fields than in others; and the ways of dealing with conflict can also be expected to vary from field to field. In the area of general organizational theory, Charles Perrow's (1984) analysis of "normal accidents" provides a useful framework for addressing social-structural variations in conflict emergence and resolution. I shall briefly present the core of Perrow's argument, and then illustrate how our rudimentary organizational model of science might explain scientific controversies.

Perrow (1984) is interested in explaining why "normal" accidents occur in organizations employing highly advanced and complex technologies, such as nuclear power plants. Normal accidents are those that are built into the very structures of complex technological systems, and hence cannot generally be prevented by improved organizational designs, more complete information, or a more skilled organizational staff. Perrow's rather gloomy diagnosis is that disaster is the "normal" price we must pay for highly complex technological systems.

Normal accidents are likely to occur in organizations with technologies that combine "complex interactions" with "tight coupling" of systemic processes. Complex technological systems consist of dynamically interrelated instead of segregated subsystems, and they arrange their production steps in close physical proximity. Hence, in complex systems technical failures are more likely to trigger surprising and unpredictable feedback effects than in linear systems where production processes are arranged sequentially. As a consequence, complex systems are generally less well understood than linear systems, and technical problems are much more difficult to locate and resolve. Whereas linear systems with serial interdependence of production steps—such as assembly line production—establish highly visible causal chains and generate more predictable work outcomes, complex technosystems with reciprocal interdependence of production steps—such as nuclear power or petrochemical plants—are much more susceptible to unpredictable disturbances rapidly ramifying throughout the system.

According to Perrow (1984), combining complex technological interactions with "tight coupling" of production processes yields the formula for normal accidents and disaster. In tightly coupled technosystems, disturbances cannot easily be isolated, and there are only a few alternative ways of operating the system in case some part or process fails. If something goes wrong in tightly coupled systems, the problem will quickly and unpredictably affect other parts of the system as well. On the other hand, in loosely coupled technosystems, such as pooled production plants, alternative operating methods are more readily available. In such systems problems can be isolated and worked on while other functionally and spatially segregated parts may continue to operate normally.

We can easily translate Perrow's central variables, the degree of technological complexity, and the tightness of coupling between subparts and subsystems, into our familiar organizational variables of task uncertainty and mutual dependence. The more complex the technology, the less predictable its causal interactions, and the more uncertain the work process. Similarly, the tighter the coupling between systemic parts and processes, the higher the overall level of mutual dependence invested in a production system. How can this organizational model be applied to explain scientific controversies as normal accidents?

Figure 4.4 arranges four kinds of scientific change according to the internal organization of various scientific fields. A more complete treatment of change will be given in chapter 6.

Specialization (Box A). Competition will lead to specialization under conditions of low dependence and uncertainty. Specialization refers to the more routine extensions of established paradigms to diverse problem areas. What is changing is not so much knowledge itself, but rather its application to new problems. If mutual dependence is low, there is room for distinct subspecialties to establish their own reputational organizations. There is not a great deal of paradigmatic unity in the overall field, but there are distinct clusters of practitioners who are organized around specific subject matters. Change is brought about through refining and specifying research on distinct areas of reality. Such specialties typically tackle fairly narrow questions and pursue piecemeal improvements in knowledge and technical sophistication in methods without dramatic innovation. This appears to be the change pattern within the numerous subspecialties in sociology, such as demography, criminology, or medical sociology.

Alternations between Normal and Revolutionary Science (Box B). Under these conditions, competition will not so much lead to pro-

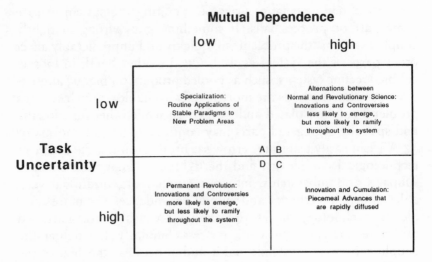

FIGURE 4.4   A Model of Scientific Change

nounced specialization, for mutual dependence is higher and, as a result, the collective controls over practitioners are tighter. There is less room for highly independent specialty clusters that hardly take notice of research in other specialties. To be sure, there is some amount of specialization in such fields, but the overall field is more unified and paradigmatically integrated. This situation resembles Kuhn's (1970) description of normal science with high paradigmatic coherence and routinized puzzle solving.

But high dependence and integration also mean that the major competitive innovations that do occur are much more likely to reverberate throughout the entire field than in disciplines where dependence is so low that specialties proceed largely independently from one another. In these rare cases, there may be revolutionary paradigm shifts that change the cultural outlook of entire fields. This situation corresponds to closely coupled technosystems in which disturbances cannot be isolated, but ramify quickly throughout the entire system. But such revolutionary upheavals are rare, and there is always the alternative of migration. This latter pattern has been demonstrated for the case of radio astronomy (Edge and Mulkay 1976; Mulkay 1975), and for the emergence of experimental psychology in nineteenth century Germany (Ben-David and Collins 1966).

Innovation and Cumulation (Box C). The condition of high dependence and uncertainty characterizes the innovative core areas of rapidly changing scientific fields. These are the elite research frontier groups working on the cutting edge of the field and defining where the field as a whole is moving. Competition is extremely intense here, there is a good deal of secrecy, and information is exchanged informally through personal networks and circulating preprints. Priority conflicts are most likely to occur in fields like these. Due to the high uncertainty involved in tackling innovative core problems, practitioners enjoy quite a bit of discretion, but are, at the same time, members of very densely organized and exclusive networks.

Change takes place here as rapid cumulation. Knowledge is quickly outdated, and practitioners are under a lot of pressure to keep up with the most recent advances. Traweek (1988) has described this situation for high-energy physics, but we would expect it wherever there is an elite core group of scientists pushing the frontiers of scientific knowledge. It seems to be this type of scientific change that philosophical realism had in mind when celebrating the gradual approximation of knowledge toward Truth.

Permanent Revolution (Box D). When high uncertainty is coupled with weak internal organization, the overall field will likely experience fragmentation into a variety of groups and perspectives. The *overall* level of uncertainty is high, for there are many theories and ways of doing research. Competition will not lead to cumulation, but rather to cognitive disorganization, for the collective pressures holding the field together are very weak. Since task uncertainty is high, innovations, even radical ones, do occur frequently, but are often indistinguishable from fads. Innovations are not likely to spread throughout the entire field, for the distinct schools and approaches are only loosely coupled. Hence, innovations will be contained within the separate subspecialties, and are not very likely to affect those working outside the innovating specialty.

In sociology, for example, recent innovations, such as "feminist sociology" or the "sociology of emotions," are frequently organized as distinct subspecialties that create their own communication systems and reputational networks, possibly even specialized publication outlets. Such internal differentiation and fragmentation has been the historical pattern of sociology's disciplinary development at least since World War II (Turner and Turner 1990). If novel research programs and creative innovations can be sustained by separate organizational networks within a field—such as largely autonomous and independent regional and specialty associations, journals, and professional meetings—innovations are more likely to be contained within subfields so as to not affect an entire discipline. As Jon Turner (personal communication) remarks, even if sociology had its Newton or Einstein, he or she would probably create a specialty association instead of triggering a paradigmatic revolution transforming the entire field.

Although this model of change is incomplete and schematic, I believe it does illustrate that change processes are not uniform and stable but covary with social structure. Change is a "normal accident" in Perrow's (1984) sense, for competition always drives scientists toward novelty and innovation. But Perrow also shows that not all technosystems are equally prone to normal accidents and, likewise, the *type* of change triggered by competition depends on how the relevant groups are socially organized. This view no longer assumes, with Kuhn, that it is ultimately empirical reality that drives change, nor, with the sociology of scientific knowledge and EPOR, that some unspecified and uniform social factors govern change.

SUMMARY AND CONCLUSION

My opening question for the present chapter was: How social are social studies of science? The answer, I believe, must be negative. Social factors are still conceptualized in the way realist epistemology conceptualized the external factors intruding into the internal workings of science. But a strong social theory of scientific knowledge cannot consider its work done when it has shown time and again that the standard philosophical model of rational science cannot account for what is actually happening in science. A strong social theory of science must at least meet these three basic requirements: it must show how the very core processes of science, such as fact production, are social processes; it must have a strong explanatory orientation (which includes the reflexive application to sociology itself); and it must provide a comparative framework for assessing historical and disciplinary variations in scientific practices.

Organizational and neo-Durkheimian theories are able to fulfill these requirements. When allowed to vary, major concepts and findings in microsocial studies of science—the idiosyncratic nature of scientific reasoning, the dynamics of fact production, and controversies—can be shown to covary with social-structural patterns of scientific organizations. The underlying argument is very materialistic and Durkheimian: group structures shape collective cognitions and cognitive practices, whether in "ordinary" social groups or in scientific communities.

Of course, I realize that the organizational model introduced so far is rather schematic, incomplete, and oversimplifies the actual diversity found in scientific fields. In particular, I have not yet accounted for *internal* variations within social and natural sciences. Moreover, in its present form the model is not complex enough to cover the actual organizational and cognitive dynamics of scientific fields. To develop a more fine-tuned and comprehensive theory of scientific organizations will be my major task in the remainder of this book.

CHAPTER 5

The Technological Paradigm in Organizational Theory

So far, I have used the "technological" paradigm in organization studies more or less implicitly in my critical reconstructions of microsocial studies of science. In this chapter, I shall examine that paradigm more closely, for it is the basis of the general theory of scientific organizations. Also, the concepts of "technology" and "structure" have yet to be introduced in a more elaborate and systematic fashion. To these ends, I shall present a few classical studies in organizational technology and structure. In doing so, some empirical and conceptual problems of technological theory will be discussed. These analyses will lead into a comprehensive technological model of organizational structure serving as the starting point for my discussion of scientific organizations. The chapter closes reviewing current debates in organizational theory. It appears that these debates are strikingly similar to those in the philosophy and sociology of science. This suggests once more that science and organization, or cognition and structure, are deeply interrelated.

WOODWARD'S STRUCTURAL TYPES OF TECHNOLOGY AND ORGANIZATION

Joan Woodward's (1965/1980) classical study of one hundred manufacturing firms in South East Essex was conducted at a time when neoclassical administration theory and the human relations school dominated the field of organization and management studies. The new technological approach introduced by Woodward was triggered by her observation that commercially successful firms had very different structures; a finding that was in obvious conflict with the neoclassical insistence on scientific principles prescribing

111

one best way to organize. Woodward came to the conclusion that organizational structures covaried with the types of work done in an organization so that there was no one best way to organize. That is, commercially successful firms appeared to "match" their technologies and structures, whereas less successful firms had structures that were "inappropriate" for the type of work done.[1]

The underlying assumption of technological theory is very materialistic: the type of work done in an organization, the nature of the tools and techniques used, and the shape of the raw materials determine the ways in which task-related interactions between organizational members are socially structured. However, Woodward (1965/1980:11) had very vague and inconsistent notions of technology and structure: technology comprised such diverse aspects as production density and product diversity, the time spans of work operations, the degree of automation, and the flexibility of production facilities. Her notion of structure (1965/1980:12f.) was similarly imprecise: structure comprised the formal authority and decision hierarchies laid out in organizational charts; the cost ratios of wages, materials, and administrative overheads; the proportions of various groups of organizational personnel; the number of levels in the command hierarchy; and the size of control spans.[2]

Woodward's most important contribution to the technological school is her distinction between three technological and structural types of organizations (1965/1980:37–40). *Unit and small batch* production plants manufacture goods or deliver services according to the special requirements of individual clients and customers. Unit production plants typically manufacture one or a few unique goods at a time so that work processes and task environments vary frequently. Examples are traditional crafts such as custom-made tailoring, the movie industry, or scientific laboratories.

Large batch and mass production plants manufacture larger varieties and quantities of goods according to more standardized and mechanized procedures. Products are assembled from various parts, and the coordination of the workflow is achieved through the organizational structure. Examples are the assembly line production of cars or highly routinized people-processing institutions such as elementary schools.

Process or continuous flow production plants coordinate the workflow through control mechanisms built into the very structure of the technology itself. Usually, some raw material is gradually transformed into a final product through a highly mechanized or even automated series of closely coupled operations. Examples are oil refineries, drug manufacturing, and most foods.

Woodward (1965/1980) appears to believe that these production types represent increasing levels of technical complexity, with unit production being the least and process production the most "technically advanced" (p.51) systems. However, neither the conceptual nor the operational rationales behind this scaling are convincing; and it is precisely the group of her findings associated with this technical complexity variable that has drawn the most empirically based objections against Woodward's work (e.g., Blau, McHugh, McKinley, and Tracy 1976). According to Woodward, the technical complexity variable is responsible for the *linear* relationships she found between technology and structure. Technical complexity increases the number of hierarchical levels in an organization (p.51); increases the control spans of chief executives (pp.52ff.); decreases the proportion of wages and salaries to total organizational expenditures (p.55); and increases the number of graduates employed by an organization (p.57).

The linear relationship between technology and structure, however, has turned out to be less important for subsequent developments in technological theory than the more conclusive *curvilinear* relationship Woodward found. Also, this curvilinear relationship has been replicated more reliably than the direct relationship (Lincoln, Hannada, and McBride 1986:347). According to Woodward (pp.60ff.), unit and process production plants share certain structural characteristics that are absent in large batch plants. Using Burns and Stalker's (1961) distinction between "organic" and "mechanistic" systems, Woodward found that unit and process production plants have more informal communication patterns, less organizational rigidity, and more flexible role definitions than large batch or mass production plants. As organic systems, unit and process production plants coordinate the workflow more through personal and mutual consultation than the mechanistic systems of mass production plants which rely more on formal command lines and a great deal of paperwork. Mechanistic systems use more formal and elaborate controls and sanctions to secure compliance, whereas organic systems are more participatory and allow for more worker discretion.

Due to her imprecise notions of technology and structure, Woodward's own rationale for the curvilinear relationships between these two variables is not very conclusive. I suggest that the curvilinear relationship between technology and structure reflects the levels of task uncertainty and forms of interdependence found in Woodward's production types. Unit and process production plants are informal organic systems because their levels of task uncertainty are higher than in mass production, and because their workflows require more flexible forms of coordination. In unit production, task uncertainty is generally high because unique products require skilled labor and nonroutine task performance. Hence, worker discretion must be fairly high, and coordination is not likely to be achieved by formal command lines and regulations.

In process production, the coordination of the workflow is built into the technology itself, but due to close coupling, uncertainty is high because disturbances quickly ramify throughout the system (Perrow 1984). Hence, in process production highly skilled maintenance workers deal with nonroutine and unpredictable technical problems and, thereby, will have considerable control over their areas of uncertainty (Crozier 1964). Due to the rather complex nature of the technology, maintenance workers are more likely to resolve problems by mutual consultation than by following formal rules and regulations. This explanation of the curvilinear relationships between technology and structure actually corresponds to Woodward's own findings that report the proportion of highly skilled maintenance workers to be highest in process production, and the control spans for middle managers to be lowest there.

Conversely, mass production plants have lower task uncertainty and less tight coupling between production steps. Task and work outcomes are fairly routine, and disturbances can be isolated and worked on in a more predictable way. Consequently, these firms have more mechanistic control systems, for the workflow can be coordinated through formal rules and regulations. In my view, then, it is our familiar variables of task uncertainty and mutual dependence that explain Woodward's findings most parsimoniously. The weakest part of Woodward's study is her vague notions of technology and structure. Charles Perrow's technological theory of organizational structure represents an important advance over Woodward because it clarifies these important concepts.

PERROW'S EARLY COMPARATIVE FRAMEWORK FOR ORGANIZATION ANALYSIS

Perrow (1967) defines "technology" broadly as the work being done in an organization. More precisely, technology comprises the actions being performed on an object, and the methods and techniques employed when transforming raw materials into some final good or service. Crucial technological determinants of structure are the number of exceptions encountered during the work process, and the analyzability of the search process undertaken when exceptions actually occur. Search processes can be more logical and predictable, such as in mechanical engineering, or more uncertain and diffuse, as in psychiatric casework.

If many exceptions are combined with diffuse and unclear search procedures, the technology is "nonroutine." Conversely, the technology is "routine" when the few exceptions that do occur can be dealt with in a more predictable and straightforward way. Of course, routine versus nonroutine technologies represent the conceptual poles of an ideal-typical continuum: many combinations of the two variables (i.e., the number of exceptions and the predictability of the search processes) are empirically possible.

Perrow (1967) distinguishes technology from the raw materials being processed by an organization. Raw materials may vary on a scale ranging from "well understood" (e.g., custodial institutions) to "not well understood" (e.g., elite psychiatric counseling). Raw materials also differ in the degrees to which they are perceived as more uniform and stable or as more unstable and heterogeneous. If well-understood raw materials are perceived as uniform and stable, raw materials are routine, whereas nonroutine raw materials are not well understood and perceived as heterogeneous and unstable.

Although Perrow (1967) distinguishes between technology and raw materials, it is likely that routine technologies usually involve well-understood raw materials that are perceived as uniform and stable, whereas nonroutine technologies are typically performed on objects that are not well known, heterogeneous, and changing. Hence, I would suggest collapsing Perrow's technology and raw materials variables into the more parsimonious "task uncertainty" variable used throughout this chapter. High task uncertainty occurs when technologies are nonroutine and raw materials are unstable and not well understood, while low uncertainty occurs when well-known raw materials are being transformed by routine technologies.

Perrow (1967:195) defines organizational "structure" as the form of task-related interactions between working individuals. The two crucial aspects of structure are "control," or the amount of discretion granted to workers, and "coordination." Coordination can either be achieved by formal rules and programs ("coordination through planning"), or by negotiations and mutual consultation ("coordination through feedback"). Perrow (1967:199) argues that by depending on various combinations of technological and structural variables, organizations roughly belong to one of four ideal-typical structural groups. This typology is reproduced in figure 5.1. I shall focus my discussion on cells 2 and 4. Note how similar this typology is to the rudimentary technological model of scientific organizations I have used to reconstruct microsocial studies of science.[3] Cell 2 represents organizations that apply nonroutine technologies to raw materials that are not well understood and perceived as rather uncertain and unstable (i.e., high task uncertainty). Organizations dealing with such technologies and raw materials generally have rather informal, flexible, and "polycentral" structures that achieve coordination more by mutual consultation than by centralized planning.

Conversely, cell 4 contains organizations that employ routine technologies to transform more predictable and stable raw materials (i.e., low task uncertainty). Such organizations need not grant workers much discretion and control over their task areas, and hence may integrate the workflow by more formal programs and centralized command hierarchies. That is, these organizations display what Burns and Stalker (1961) and Woodward (1965/1980) call "mechanistic" systems, while those organizations corresponding to the structural type displayed in cell 2 are "organic" systems.

Perrow's (1967) model provides much clearer definitions of technology and structure, and it also offers more conclusive ration-

	Discre-tion	Power	Coord. w/in gp.	Interde-pendence of groups	Discre-tion	Power	Coord. w/in gp.	Interde-pendence of groups
Technical	low	low	plan		high	high	feed	
				low				high
Superv.	high	high	feed		high	high	feed	
	Decentralized				Flexible, Polycentralized			
				1	2			
Technical				4	3			
	low	high	plan		high	high	feed	
				low				low
Superv.	low	low	plan		low	low	plan	
	Formal, Centralized				Flexible, Centralized			

FIGURE 5.1   Perrow's Early Typology of Organizational
Structures
From Charles Perrow, "A Framework for the Comparative Analysis of Organizations"
American Sociological Review 32:194–208, 1967, Figure 3. Reprinted with permission.

ales for the links between these variables than Woodward's model did. However, in his early work, Perrow (1967) did not yet clearly distinguish between technological and structural aspects of the interdependencies built into an organization's workflow. That is, the early Perrow did not yet differentiate clearly between the coordination requirements imposed by the technology itself, and the structural arrangements organizations establish to cope with these requirements. James Thompson's (1967) approach offers a more fine-tuned analysis of these subtle, yet important, distinctions. I shall briefly introduce his approach and then return to the more elaborate contingency model suggested by the later Perrow.

THOMPSON'S TECHNOLOGICAL INTERDEPENDENCE TYPES

James Thompson (1967:54ff.) distinguishes between three types of technological interdependence. *Pooled* interdependence describes a situation where organizational subunits only depend on the larger organization, but not on each other, in their operations. Typically, pooled interdependence is found when subunits or branches participate in a resource pool provided by the overall organization, but do their work largely independently from each other. Examples are the DMV, post offices, and most prisons.

Sequential interdependence exists when production steps are arranged serially. In this case, the output of one subunit or branch is the input for the subsequent subunit or branch, but not vice versa. The most prominent examples are assembly line production or highly routinized people-processing institutions such as elementary schools with rigidly programmed course sequences.

Reciprocal interdependence is given when subunits depend closely on each others' operations. In this case, the inputs of one subunit are produced as outputs by another subunit, and vice versa. Examples are scientific research and the maintenance and production divisions in aircraft manufacturing.

Thompson's (1967) main point is that these various technological forms of interdependence require various corresponding patterns of coordination. Pooled interdependence is compatible with highly standardized and hierarchical coordination patterns, for coordination mostly involves the distribution of resources to largely independent subunits. Coordination can be achieved by *formal rules and regulations*, for pooled interdependence primarily requires highly repetitive and predictable decisions on budget allocation.

Sequentially interdependent subunits, on the other hand, are more likely to be coordinated by *plans and schedules*. Sequential interdependence is somewhat more complex and vulnerable to disturbances so that coordination must, to an extent, be open to unforeseeable contingencies. And reciprocally interdependent subunits are likely to be coordinated by *mutual adjustments* and consultation. Since subunits are very closely coupled and usually involve complex technological feedbacks, coordination cannot be accomplished by standard rules or programming. Rather, subunits are given considerable leverage and discretion for flexibly deciding on mutually relevant issues.

PERROW'S LATER COMPARATIVE FRAMEWORK FOR ORGANIZATION ANALYSIS

Two significant advantages of Thompson's (1967) technological interdependence typology are that he adds one more structural type of organization, "pooled" production, to Woodward's (1965/1980) original schema, and that he clearly distinguishes between *technological* forms of interdependence and *structural* forms of coordination. The later Perrow (1984) adopts this clear distinction in his treatment of the "normal" accidents that are built into the very technosystems of complex organizations. Since I have introduced Perrow's argument in more detail in my discussion of scientific change, I shall only very briefly recapitulate his main points here.

Perrow's (1984) two main variables for analyzing organizational technosystems are the complexity of interactions and the tightness of coupling between subunits of the workflow. Complex interactive systems involve what the early Perrow (1967) calls "coordination through feedback," and what Thompson (1967) calls "reciprocal interdependence" between subunits. Since complex interactive systems are generally not well understood and require nonroutine work techniques, I have suggested that complex interactive systems are those with high levels of task uncertainty.

Conversely, linear systems involve what the early Perrow (1967) calls "coordination through planning," and what Thompson (1967) refers to as "sequential" interdependence. Since linear systems are generally well understood and can be handled by more routine work techniques, I have suggested that linear systems are those with comparatively low levels of task uncertainty.

While the interaction variable specifies the *type* of interdependence built into a technosystem, Perrow's (1984) second variable, the tightness of coupling between subunits, refers to the *degree* of interdependence between subunits. Tightly coupled technosystems do not allow for delays in processing, arrange production steps according to largely invariant sequences, and do not dispose over many alternative ways of operating the system. I have suggested that tight coupling covers much the same dimensions of technology as high mutual dependence. Conversely, loosely coupled systems permit the isolation of disturbances, have more flexible production schedules, and more alternative ways of operating the system. It appears that loose coupling corresponds to low mutual dependence.

Perrow (1984) uses four possible combinations of his coupling and interaction variables to predict four structural types of organizations that "match" their technologies. This typology is reproduced in figure 5.2. I have replaced his labels of coupling and interaction with mutual dependence and task uncertainty.[4]

Since the later Perrow's model corresponds even more closely to the model I have been using, I shall focus on cells 1 and 3. Under conditions of low uncertainty and high dependence (cell 1), organizational structures can be more centralized and formalized, for the technology is more routine, and close mutual dependence requires tight structural coordination between subunits. When task uncertainty is low, organizational practices will be more standardized and structured according to more formal rules and regulations. Worker discretion will be low, and hence coordination will be achieved through formal command hierarchies.

Conversely, under conditions of high uncertainty and low dependence (cell 3), organizational structures will be more decentralized and informal. Tasks are less routine and predictable, and hence work will be governed by more informal rules and idiosyncratic directions. Low mutual dependence implies that coordination problems will not be very severe so that worker discretion will be rather high. Under these conditions, control struc-

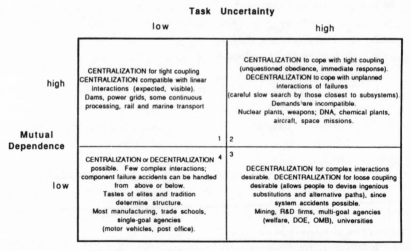

FIGURE 5.2   The Later Perrow's Typology of Organizational Structures

From *Normal Accidents: Living with High-Risk Technologies*, by Charles Perrow. Copyright © 1984 by Basic Books, Inc. Reprinted with permission of Basic Books, a division of Harper Collins Publishers.

tures will be more decentralized since workers will have considerable autonomy in performing their jobs.

So far, my review of the technological tradition in organizational theory has gradually moved toward a more comprehensive and conclusive theory of organizational structure. Starting with Woodward's (1965/1980) structural types, we have seen how Perrow's (1967) early model clarified the notions of technology and structure and offered more convincing rationales for the relationships between these two variables. Thompson's (1967) technological interdependence types further clarified the distinctions between technological forms of interdependence and structural forms of coordination. These more precise distinctions are adopted in Perrow's (1984) later technological model of organizational structure, which corresponds rather closely to the model of scientific organizations I have been using throughout the previous chapter.

But this model is still incomplete. One of the most important conceptual criticisms of contingency theory has been that as a "closed systems" approach, it neglects environmental determinants of structure (see Dawson and Wedderburn 1980; Scott 1981). Recently, the environment has assumed a prominent position in organization studies; be it as a resource environment (Pfeffer and Salancik 1978), as a selective and competitive market environment (Hannan and Freeman 1977), or as the cultural and institutional environment of organizations (Meyer and Scott 1983). I believe the environment can be included in a comprehensive technological theory of organizations, and Lawrence and Lorsch's (1967) study of organizations in various environments illustrates just how such an integration might be accomplished.

LAWRENCE AND LORSCH'S ENVIRONMENTAL MODEL OF ORGANIZATIONAL STRUCTURE

Like many other contingency approaches to organizational structure, Lawrence and Lorsch's (1967) study of organizations in diverse environments was motivated by dissatisfaction with the neoclassical insistence on one best way to organize. Like Woodward (1965/1980), Lawrence and Lorsch observed that commercially successful firms in different environments had different structures so that following the scientific administration principles of neoclassical theory could hardly be sound advice to managers. But unlike most technological approaches to organizational structure, Lawrence and Lorsch (1967) conceive of organizations as open systems reacting to various environmental influences. Most technological theories view organizations instead as closed systems which are solely determined by the structural demands of technological throughputs.

A further advantage of their model is that environment, technology, and structure are not conceptualized as uniform entities, as if organizations had only one structure reacting to only one type of technology and environment. Rather, Lawrence and Lorsch (1967) acknowledge that different divisions or departments within the same organization might process different technologies, react to different segments of the environment, and hence might develop different internal structures and practices. It might just be that the inconclusive and controversial empirical evidence gathered by tests of technological models is partly due to the frequent failure to account for *internal* variations in organizational technologies and structures.[5]

Lawrence and Lorsch (1967) selected ten organizations for an in-depth study of environmental impacts on structure. Environments are assumed to differ on two crucial variables: the degree of heterogeneity, and the amount of change. Heterogeneous environments are those that place a variety of adaptive demands on organizations, such as a broad diversity of customers and clients, multiple and fragmented resource pools, and a large number of other organizations that either compete with or regulate the focal organization. Conversely, homogeneous environments consist of more uniform groups of customers and clients, less diverse resource pools, and fewer competitive and regulatory organizations.

Lawrence and Lorsch (1967) collapsed these two environmental variables of heterogeneity and change into their "subjectively perceived uncertainty scale" on which middle and upper-level managers were asked to rate the environments of their organizations.[6] Presumably, the higher the rate of change and the more heterogeneous the environment, the higher the amount of perceived uncertainty. Lawrence and Lorsch's scale measures uncertainty by the clarity of information managers have about the organizational environment, by the degree of cognitive ambiguity in assessments of causal relationships between environmental variables, and by the time it takes for organizational actions to trigger relevant environmental feedbacks. Perceived uncertainty is high when information about the environment is unclear and imprecise, when there is a great deal of cognitive ambiguity, and when organizational actions trigger environmental feedback effects that arrive too late to clearly assign specific actions to specific environmental reactions.

Lawrence and Lorsch (1967) assigned their ten organizations to three different types of environments. Six plastics firms operated in environments with a high overall degree of perceived uncertainty, two food corporations dealt with environments perceived as medium uncertain, and two container firms had low overall degrees of perceived environmental uncertainty. The basic result of Lawrence and Lorsch's study is that in all three industries, those organizations that "matched" their structures with their perceived environmental conditions were the most efficient and commercially successful.

Generally, the plastics industry operates in a heterogeneous and rapidly changing environment that is perceived as highly uncertain. Commercially successful firms in the plastics industry are highly differentiated into multiple departments that deal with various segments of the heterogeneous environment. These organizations had decentralized control structures, highly informal and flexible work practices, and accomplished integration by mutual adjustments and cooperation rather than by formal command hierarchies. Heterogeneous and rapidly changing environments appear to necessitate more internal structural differentiation and, due to high uncertainty, establish more worker discretion and flatter hierarchies.

Conversely, firms in the container industry deal with more homogeneous and stable environments that are perceived as rather predictable and certain. Hence, the most commercially successful firms are those with comparatively low levels of structural differentiation. In these firms, the workflow is coordinated through a formal command hierarchy that relies more on codified rules and regulations than on informal reciprocal adjustments. Due to the routine character of the work, worker discretion is rather low, and the overall organizational structure was found to be more centralized than in the plastics industry.

And finally, commercially successful firms in the food industry operate in an environment of intermediate uncertainty, and thus have structures that represent a combination of those found in the plastics and container industries. In sum, commercially successful firms coalign their internal structures with the conditions found in their environments. The higher the degrees of heterogeneity, change, and perceived uncertainty, the higher the level of internal structural differentiation, and the more informal and decentralized the patterns of coordination.

Lawrence and Lorsch (1967) obtained similar results upon comparing different departments *within* the same organization. For example, the research, market and sales, and production departments of plastics firms all deal with different segments of the environment. The research department perceives its environment as much more uncertain than the other departments, with the production department operating in the most certain environment. Amongst other results,[7] Lawrence and Lorsch (1967) found that departments facing highly uncertain environments, such as research, have less formal control and coordination structures, whereas departments coping with low uncertainty, such as production, have taller hierarchies, more formal work standards, and higher ratios of supervisors. Generally, the higher the uncertainty departments have to cope with, the more informal and decentralized their coordination and control structures.

Lawrence and Lorsch's (1967) classic study corroborates the importance of "task uncertainty" as a crucial variable shaping organizational structure. Our review of the technological tradition in organizational theory suggests that together with the type and degree of mutual dependence, task uncertainty is the common conceptual denominator of most technological theories of organizational structure (Stinchcombe 1990:97).[8] Lawrence and Lorsch have extended the internalist perspective predominant in technological approaches to include environmental sources of uncertainty and complexity. Moreover, their two environmental variables of heterogeneity and change—aggregated into the perceived uncertainty scale—appear to cover much the same aspects most

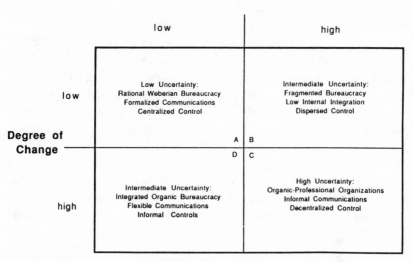

Degree of Heterogeneity

FIGURE 5.3   Environmental Impacts on Organizational Structure

"informational" approaches to the environment hold to be important in shaping organizational structure (Aldrich and Mindlin 1978).[9]

Figure 5.3 summarizes this line of argument. Note how parsimoniously task uncertainty covers both technological and environmental sources of structural complexity: the *environmental* predictions about organizational structure given in figure 5.3 are almost identical with the *technological* predictions summarized in figures 5.1 and 5.2.

CONTROL THEORY

Organizational control theory (Etzioni 1975; Collins 1988) has argued that organizations differ in the control devices they use to secure members' compliance. Although originally introduced as an alternative way of classifying organizations, control theory can be integrated into a technological theory of organizations. To complete my review of the technological paradigm in organizational theory, I shall briefly outline how control devices covary with organizational technologies. This discussion will lead into a comprehensive model of organizational structure that will serve as the starting point for the analysis of scientific organizations.

Etzioni (1975) distinguishes between three control devices organizations may employ to secure behavioral compliance. Of course, empirical organizations often employ combinations of several control devices so that Etzioni's schema has the conceptual status of an ideal typology. *Coercion* is the typical control device employed by preindustrial organizations, although modern armies, for example, also rely to a large extent on coercion and formal sanctions, at least during peacetime. Coercion generates high social and emotional costs, for it leads to low member motivation and alienation, sometimes even to overt resistance and conflict.

Material rewards are used by almost all contemporary organizations, except for voluntary associations and social movements. Material rewards lead to instrumental orientations among organizational members. Employers and employees organize as bargaining coalitions negotiating salaries and wages. Although less alienating than pure coercion, material rewards cannot be expected to motivate members to perform at levels above and beyond what is minimally required by labor contracts and agreements.

Normative control devices are most often found in voluntary associations and social movements, and generally in all organizations lacking substantial property resources. Normative control occurs when members have internalized organizational goals and values, and hence are driven by high intrinsic motivations to perform at levels exceeding what can be expected in exchange for material rewards alone. Normative control may be accomplished by extended organizational socialization, by an organization's ritual and solidarity-generating informal activities, or by actual or anticipated opportunities to participate in leadership and decision making.

Control types correspond to particular administrative devices (R. Collins 1975). Coercion requires permanent *surveillance*, a large staff of supervisory personnel, and a large amount of resources to enforce frequent negative sanctions. Material rewards are distributed on the basis of *performance assessments*. To evaluate and reward performance, a large administrative staff is needed to process and store files. Hence, material rewards generate rather bureaucratic forms of administration involving a great deal of paperwork. Normative control is exercized as control over *information* and definitions of reality (Crozier 1964). Those members exercizing normative control define the informational premises upon which organizational decisions are being made and implemented (Simon 1945/1976). Control over information cannot be widely dispersed without eroding the organizational power structure, for multiple and possibly conflicting definitions of organizational realities impede an organization's ability to decide and act.

From a technological perspective, control types and administrative devices correspond to the technological throughputs of an organization. Coercion and surveillance might be employed when tasks require only minimum initiative on the part of workers, and when task outcomes are highly visible and predictable. Total institutions such as prisons might supplement coercion and supervision by "environmental" control of inmates. That is, the sheer geography of prison cells, walls, and corridors controls behavior. Environmental control is also found in process production industries where the very technology built into the workflow coordinates activities.

Similarly, material rewards are distributed for activities yielding highly visible and predictable results. Since material rewards are allocated on the basis of work outcome inspections, the tasks involved must be rather routine and comparable across various activities. Outcomes cannot be assessed in terms of their monetary value unless tasks are fairly certain and repetitive.

Tasks involving high uncertainty, rather unpredictable outcomes, and nonroutine skills, on the other hand, cannot easily be controlled by coercion and material rewards alone, for these tasks necessitate a good deal of worker discretion and initiative. Normative control devices will be employed whenever purely instrumental orientations are insufficient to get the work done, and when highly unpredictable work outcomes prohibit the use of standard work inspection and assessment methods. For example, scientists are often given considerable control over their areas of uncertainty, and are expected to gain satisfaction from the pursuit of knowledge itself, not just from their material rewards. This explains the Mertonian preoccupation with the institutional norms of science. But normative control is in no way *constitutive* of science; rather, it

Production Types	Environment	Task Uncertainty	Mutual Dep.	Coord. Patterns	Control Forms	Admin. Devices	Organizational Structure
Unit (Movies, Labs)	changing, heterogeneous	high	reciprocal, high	mutual adjustments consultation	normative	information, knowledge	informal, decentral
Large Batch (Mass Prod.)	changing, homogeneous	medium to certain	serial, linear, low	plans, schedules	material rewards	outcome inspection, rules	mechanistic bureaucracy
Process (Oil Refinery)	changing, homogeneous	medium to uncertain	serial, complex, high	technological	material rewards	environmental	organic bureaucracy
Pooled (Prisons)	stable, homogeneous	certain	pooled, low	command hierarchy	coercion	supervision	formal, centralized

FIGURE 5.4   A Technological Model of Organizations

is used whenever work requires high initiative and commitment on the part of the workers.

Control types and administrative devices covary with the degrees of task uncertainty inherent in various types of organizational technologies. The more uncertain the task, and the less comparable and predictable the work outcomes, the more worker discretion and autonomy will be required, and the stronger the emphasis on internalized normative controls. Conversely, the more routine the tasks, and the more predictable the work outcomes, the more likely it is that work activities will be rewarded monetarily or controlled coercively.

This concludes my review of the technological paradigm in organizational theory. Although there are some unresolved conceptual and empirical problems, I believe technological theory remains the most comprehensive approach currently available in organizational theory. Stinchcombe's (1990) "informational" theory shares this assessment. Most importantly, technological theory has a strong comparative and explanatory orientation. It explains variations in organizational structures, in the possible rationality of organizational behaviors, in control types, and in administrative devices. To reconstruct the explanatory strengths of technological theory was my goal in the present section, for that theory will guide my analysis of scientific organizations.

Figure 5.4 summarizes and synthesizes the technological argument; combining Woodward's production types, Perrow's complexity and coordination variables, Lawrence and Lorsch's environmental dimensions, Thompson's types of technological interdependence, and Etzioni's control forms. Of course, empirical organizations are likely to employ some combination of these variables; they will not appear in the pure forms presented here. Likewise, the differences might be between units inside organizations, not necessarily between entire organizations.

CURRENT DEBATES IN ORGANIZATIONAL THEORY

The theory of scientific organizations has the strong implication that predictions about organizational structures and practices do not differ significantly for scientific and nonscientific organizations. The review of the technological tradition suggests that these predictions are indeed fairly stable. Organizations and departments have more formal and rigid structures when they deal with low uncertainty, and they have more informal and decentralized structures when facing uncertain tasks. In the present section, I shall pursue this implication further in examining contemporary debates in organizational theory. These debates revolve around the problem of rationality in organizations, and they are fought by two opposing camps. The Weberian and neoclassical tradition views organizations as rational goal-achieving instruments, while garbage-can and institutionalist approaches view rationality as more of a myth. There are striking and very instructive resemblances between the debates in organizational theory and in the philosophy and sociology of science. Organizational rationalism and epistemological realism view organizations and science as orderly and systematic, while garbage-can models and social studies of science argue that organizations and science are more disorderly and idiosyncratic. Also, rationalism/realism and garbage-can models/social studies of science share the same cognitive habitus. It turns out that in both cases, debates are based upon the same conceptual fallacy: not allowing for variations in organizational and scientific practices. The technological argument avoids this fallacy by suggesting that organizations and science will be rational when they can be, but will be more disorderly when facing high task uncertainty.

Ever since Weber's (1922/1947) classic theory of bureaucracy, a most controversial issue in organizational theory has been the *rationality* of formal organizations. In fact, the various schools and paradigms in contemporary organizational theory can be defined and distinguished by the ways in which they approach the problem of organizational rationality. Weber's strong claim was that bureaucracy was the most rational and efficient form of administration in complex modern societies, but contemporary organizational theorists disagree rather markedly in their assessments of the role rationality plays in shaping actual organizational structures and behaviors. The cleavages in current organizational theory run so deep that some scholars continue to view organizations as purposefully designed and rational instruments for collective goal achievement, while others believe that rationality is an ideological myth organizational elites create to legitimize their practices to the general public.

After Weber, the view that organizations are first and foremost rational systems has most strongly been defended by what has come to be known as the "neo-Weberian" or "neoclassical" administration theory of scholars pragmatically interested in improving management—scholars such as Gulick, Urwick, Fayol, or Taylor (see Perrow 1972; Scott 1981 for overviews). Neoclassical administration theory regards organizations as purposefully designed instruments to effectively and efficiently achieve social goals. The role of administration theory is to design scientific principles of rational management. "Rationality" is understood in Weber's sense of instrumental means-ends relationships: a rational organization is one that achieves its goals with optimum ratios of gains to investments.

To maximize organizational rationality, management must follow the principles of scientific administration. Neoclassical theory has a strong normative and prescriptive orientation, for the principles of rational management are not observed empirical regularities but recommendations for improving organizational performance. Among these principles are unity of command and a fixed level of control spans. That is, increasing organizational performance generally requires subordinates to receive orders from one superior only, and superiors to control only a limited number of employees. Neoclassical administration theory holds that rational organizations must distribute specialized positions according to expertise and achievement, which in turn necessitates a rather centralized organizational structure with formal mechanisms of coordination. Rules and regulations should generally be codified so that responsibilities are clearly defined and behavior is predictable.

Neoclassical theory is based on the rational model of "economic man." Economic man maximizes profits according to rational assessments of means and ends. Implicitly, neoclassical theory assumes that the empirical knowledge required to organize is rather complete and accurate, that preferences are clearly ordered in terms of their priorities, and that there is consensus over organizational goals and performance standards. Most importantly, neoclassical theory assumes that there is one best way to organize, such as following the scientific principles of rational management. That is, all efficient and rational organizations should be structured according to the same management principles, such as specialization, control, and coordination. Neoclassical theory disregards environmental and technological differences between organizations, for the one best way to organize is some form of bureaucracy.

Neoclassical administration theory has crumbled under severe attacks from various alternative approaches. Roughly, these critical perspectives can be divided into two general lines of argument. There are those approaches that dismiss the neoclassical confidence in a normative theory of administrative principles prescribing one best and rational way to organize, but that continue to view organizations as *rational* instruments for collective goal achievement. On the other hand, there are approaches that dismiss the rationality paradigm altogether and view organizations instead as loosely coupled anarchies.

Approaches critical of neoclassical theory but loyal to the rationality paradigm suggest important modifications in the very notion of rationality. Most of these approaches relate in some way or other to Herbert Simon's influential *Administrative Behavior* (1945/1976). Simon has introduced the important concept of "bounded rationality" into organizational theory. As opposed to neoclassical theory, Simon argues that action can only be limitedly rational because actors have incomplete and often inaccurate information about means and are usually unaware of all possible and possibly more rational alternative courses of action. Under conditions of uncertain and ambiguous empirical knowledge, seemingly rational actions may have irrational consequences. Actors usually act according to unclear preferences with vague priorities. Hence, Simon replaces the fully rational "economic man" of neoclassical theory by the limitedly rational "administrative man." Viewed as corporate actors, organizations only try to satisfice, not to maximize under conditions of bounded rationality.

However, Simon (1945/1976) retains the neoclassical view of organizations as essentially rational instruments, and he also remains interested in how organizational performance can be improved. But the rationality Simon discusses is no longer the rationality of neoclassical economic actors. Rather, it is organizational designs that make rational action *structurally* possible (see also Williamson 1975). Most fundamentally, organizations account for the bounded rationality of their members by controlling their *decision premises* through authority, selective communication, and specialization. In this way, actors only have to account for limited sets of alternatives and decide on the basis of prepackaged and preselected information. Rules and specialized roles define stable expectations, and standard practices reduce the complexity of organizational decision making. Specialized positions and communication channels further reduce organizational complexity, and in this way account for the limited ability of individuals to rationally assess the available information, priorities, and value preferences. In short, organizational action can be rational because efficient structural designs reduce the complexity burdening individual decision making.

The "garbage-can" model of organization (March and Olsen 1979) pushes Simon's notion of bounded rationality to its extreme: the limits to rationality are so severe that organizational decision making is more adequately described as an *irrational* process. Under conditions of notoriously ambiguous information and highly uncertain environments, organizational decisions are only "loosely coupled" with goals and with the resulting actions (Weick 1979). Whereas traditional theory assumed that the formal structures accounted for the limited cognitive capacities of individual actors and, in this way, made rational action possible, the garbage-can model argues that formal structures themselves consist of shifting alliances and fluid participation. Therefore, formal structures do not really govern actual organizational practices, but are a way to allocate responsibilities *after* the wrong decisions have been made.

Since decisions are made under extreme uncertainty, incomplete and inaccurate information, and conflicting goals, decision outcomes are more random than planned, more idiosyncratic than systematic, and due more to ad hoc opportunities than to strategic long-term planning. As a consequence, decisions often do not resolve anything but create more problems, and no one really wants to decide on important issues anyway. Decisions are often postponed until the problem eventually disappears, and if a decision must be made instantly, sometimes no effort is made at implementing it. Under these circumstances, rational assessments of decisions are hardly more than *post festum* rationalizations of organizational action.

That is, garbage-can theorists dismiss the very idea that organizations are and can be rational instruments for collective goal achievement. As a consequence, these theorists advocate a more playful and foolish technology for organizing that values innovativeness, creativity, and flexibility more than planning, routinization, and formalistic rigidity (March and Olsen 1979; Weick 1979). Rationality is more of an organizational ideology than a reality. It is this conclusion that the most recent school in organizational theory, institutionalism, takes as its starting point.

The institutionalist school understands itself as an alternative explanation of why organizations have formal structures (Meyer and Rowan 1977). Whereas traditional organizational theory assumed that the formal structure was due to purposeful designs intended to maximize organizational performance through effective coordination and control, institutionalism agrees with garbage-can models that the formal structure does not really govern actual organizational behaviors. However, if formal structures are largely inoperative in the day-to-day workings of organizations, the problem, of course, is why organizations have formal structures to begin with.

According to institutionalist theory, the formal structure incorporates the "rational myths" prevailing in an organization's cultural environment (Meyer and Scott 1983). These myths include professional ideologies of expertise and scientific authority, widespread public beliefs and expectations concerning the rational purposes of organizations, and certain technologies that are assumed to increase performance, such as "objective" psychological methods for personnel selection, and "scientific" economic forecasting. Institutionalist theory maintains that organizations incorporate these rational myths into their formal structures to gain cultural legitimacy and to insulate their practices from critical public inspection. As opposed to previous approaches that focus on task and resource environments, institutionalism views the environment as conglomerates of *cultural* values and beliefs about "legitimate" and "rational" ways of organizing. That is, organizations do adjust to their environments, but it is cultural models of legitimate organizing that are incorporated into an organization's formal structures.

The formal and rational structure of an organization does not really determine its actual workings, but it does buffer an organization against public inspection. The economic forecasts of scientific marketing experts might actually be ignored, psychological methods for personnel recruitment might not really be observed, and the affirmative action office might have only a weak influence in recruitment matters. But, institutionalism insists, organizations do have such formal structures because the rational myths they incorporate signal to the public that all is well in an organization, that it is culturally legitimate, that it employs state-of-the-art technologies and conforms to general expectations defining how a "good" organization should be run. Actually operating the organization according to its formal structure would probably be impossible or counterproductive, but organizations need formal structures to maintain their public acceptability, while informal systems get the work done.

Institutionalist theory realizes that cultural environments have a stronger impact on public organizations than on private firms (Meyer and Rowan 1977). Whereas the efficiency demands imposed on organizations by competitive markets shape the structures of private firms, pressures for cultural legitimacy are stronger for public organizations. Consequently, institutionalism has focused empirical investigation on public organizations, most notably civil service and educational institutions (Rowan 1982; Tolbert and Zucker 1983; Meyer and Scott 1983). Generally, this line of research has found that such public organizations copy normative exemplars of very successful or publicly celebrated organizations, and that they strive for "institutional isomorphism" with their cultural environments (DiMaggio and Powell 1983). The important conceptual point here is that institutionalist theory acknowledges *variations* in the sources shaping organizational structures and their internal workings, if only variations between private firms and public institutions.

From these debates on rationality and formal structures it is clear that organizational theory has been oscillating between two extremes. Weberian and neoclassical administration theory viewed organizations as essentially rational instruments for collective goal achievement; the implication being that the formal structure actually governs organizational action. On the other pole of the continuum are garbage-can and institutionalist models that view organizational decision making as an essentially anarchic and highly unstructured process, with a rational but largely inoperative superstructure signalling cultural legitimacy.

I believe that the disputes between rationalist and irrationalist paradigms in organizational theory are based on the same conceptual fallacies as the controversy on rationality between the philosophy and sociology of science. Recall that realist epistemology and social studies of science do not account for possible variations in scientific practices. Instead, they both claim to have discovered the rational or social *nature* of science. Likewise, rationalist and irrationalist paradigms in organizational theory have a tendency to regard organizations as being either essentially rational *or* irrational, formal *or* informal, idiosyncratic *or* systematic. But instead of speculating about the rational versus irrational nature of science and other organizations, I propose again that we treat organizational structures and practices as *variables* rather than constants.

In other words, organizations will be rational when they can be, but they will be idiosyncratic and messy when they cannot be rational. Under certain conditions, organizations will actually follow their formal rules and regulations; while under different conditions, observing these rules and regulations might be impossible or counterproductive. In some cases, rationality will be what actually governs organizational practices, while in other cases, rationality will be more of a myth. There are organizations, such as the DMV or post offices, that look very much like formalistic Weberian bureaucracies, while other organizations, such as universities or the publishing industry (Levitt and Nass 1989) look more like garbage-cans. Is there a paradigm in organizational theory that is able to account for and explain such variations in organizational and behavioral structures?

In terms of the technological argument presented in the first part of this chapter, neoclassical administration theory describes organizations operating under conditions of low task uncertainty and high mutual dependence. Low uncertainty characterizes fairly predictable work outcomes, standardized work technologies, and rather routinized performance criteria. Under these conditions, work processes and technologies are comparatively simple, do not require highly skilled labor, and hence go along with low levels of worker discretion and autonomy. The critical point is that standardized work practices with predictable outcomes can be organized according to more formal rules and regulations; that is, according to the model of organizing suggested by neoclassical theory.

Especially when combined with high mutual dependence between production steps and divisions, low uncertainty will generate structures and practices that resemble a formal Weberian bureaucracy. For high mutual dependence triggers problems of coordination and control; and if task uncertainty is low, these problems are likely to be resolved by a rather centralized organizational hierarchy with formalized communication channels and well-defined intransitive authority structures. Very schematically, this condition seems to be met by mass production and assembly line plants, for highly mechanized tasks are combined here with rather severe coordination problems between the various sectors and departments (Woodward 1965/1980).

Conversely, if high uncertainty is combined with low dependence, quite different organizational structures are likely to emerge. For if uncertainty is high, work processes can hardly be

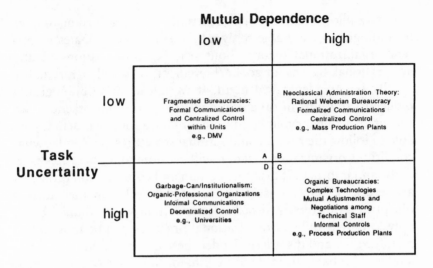

Mutual Dependence

	low	high
low	Fragmented Bureaucracies: Formal Communications and Centralized Control within Units e.g., DMV	Neoclassical Administration Theory: Rational Weberian Bureaucracy Formalized Communications Centralized Control e.g., Mass Production Plants
	A	B
	D	C
high	Garbage-Can/Institutionalism: Organic-Professional Organizations Informal Communications Decentralized Control e.g., Universities	Organic Bureaucracies: Complex Technologies Mutual Adjustments and Negotiations among Technical Staff Informal Controls e.g., Process Production Plants

Task Uncertainty

FIGURE 5.5   A Model of Organizational Theories

organized by formal and standardized rules and regulations. Highly uncertain tasks have more unpredictable outcomes that cannot easily be assessed by generalized performance criteria. Hence, highly uncertain tasks require more worker discretion and autonomy, and more informal and flexible task directions.

If combined with low levels of mutual dependence, high task uncertainty is likely to yield rather flat and decentralized hierarchies with more informal and transitive communication channels. Low mutual dependence greatly reduces problems of coordination and control so that highly centralized and formalized command hierarchies are unlikely to emerge. The condition of high uncertainty and low dependence seems to be met by what Hage (1980) has called "organic-professional" organizations—organizations that employ rather complex work technologies but whose branches or departments are only loosely coupled. Universities are an example (Perrow 1984:97). If we combine the two variables of task uncertainty and mutual dependence, we obtain the model of organizational theories depicted in figure 5.5. Of course, the model does not cover the actual variety of organizational technologies and structures, and therefore is intended here to serve heuristic purposes only.[10] The crucial conceptual point is that if we allow organizational structures to vary according to their technologies, we are able to avoid the essentialist debates in current organizational theory about whether organizations "really are" rational systems of goal achievement, or whether rationality "really is" an ideological myth. If we allow for technologically induced variations in organizational structures and practices, we can predict under what conditions organizations will actually be able to follow their formal and rational structures, and under what conditions organizational action will be more idiosyncratic, disorderly, and informal. If tasks are certain and coordination problems are severe, organizations will be more able to follow their formal blueprints. Conversely, if tasks are uncertain with subunits being only loosely coupled, organizational practices will be more informal, flexible, and disorderly. Under these latter conditions, rationality will probably be more of a myth, for the actual organizational actions and decisions will correspond more to the garbage-can metaphor of organizing.

We are now in a position to appreciate the surprising and instructive similarities between orthodox realist epistemology and the rational paradigm in organizational theory, on the one hand, and between garbage-can/institutionalist models of organization and social studies of science, on the other hand. Consider first the issue of rationality. Realism and neoclassical theory believe that scientific and organizational practices actually proceed or should proceed according to formal procedural norms of correct science and administration. In both cases, these norms are conceptualized as universal and imperative. There is one correct way to arrive at accurate representations of reality (i.e., following the rules of scientific method), and there is one best way to organize (i.e., following the principles of scientific administration). That is, both philosophical and organizational rationalism assert that all sciences and all organizations should follow the same rules of scientific method and administration, independent of all differences between them. Both belief systems do not allow for possible variations between scientific fields and organizations.

Similarly, social studies of science and garbage-can models claim that it is the *nature* of science and organizing to be idiosyncratic and disorderly. Localist models of scientific production and garbage-can models of organizing stress the unsystematic and unstructured character of scientific and organizational practices. These practices do not follow universal and rational norms of methodological propriety; rather, they draw upon contingent and contextual resources to make sense of whatever happens. Scientific and organizational activities are not guided by the firm and reliable hand of Reason; rather, they must cope with highly ambiguous information and unpredictable contingencies.

Consequently, localist models of science and garbage-can models of organizing view rationality as more of a myth than a reality. The localist model assumes scientific rationality to rest on the rhetorical conversions mundane research practices undergo when they are translated into the impartial language of Reason and Reality used in the conventional research report. Rationality is a textual fiction, not the reality of science-in-the-making. Similarly, institutionalism, following the garbage-can model, maintains that rationality is an ideological myth organizational elites fabricate to gain public approval. Rationality is an ideological fiction, not the reality of organizational action and decision making.

Realism and neoclassical theory insist that the nature of science and organizing is rational and systematic, while social studies of science and garbage-can theory insist that the nature of science and organizing is social and idiosyncratic. That is, none of these approaches is sensitive to possible variations in scientific and organizational practices. But if we do allow for such variations, we can avoid essentialist speculations about the nature of science and organizing.

Compare, for example, figure 4.1 ("A Model of Research Practices," p. 87) with figure 5.5 ("A Model of Organizational Theories," p. 137). From a technological perspective, social studies of science and garbage-can models do not describe the nature of science and organizing, but scientific and organizational practices under conditions of high task uncertainty and low mutual dependence. Under these conditions, scientific and organizational practices will be more disorderly and idiosyncratic.[11] If uncertainty is high and dependence is low, scientists and organizational actors will capitalize on locally available resources and opportunities more than on context-independent and universal rules of "proper research" and "correct decision making." In the absence of standardized work directions, the individual discretion of workers will be rather high, whereas the influence of translocal communities and centralized hierarchies will be rather low. That is, localist and garbage-can models do not describe the essential nature of science and organizing; rather, they describe one extreme pole of what should be regarded as a continuum of standardization and formalization in scientific and organizational practices.

Of course, it would be equally misleading to assume, as do the rational philosophical model of science and neoclassical administration theory, that all sciences and organizations proceed or should proceed according to the formal and universal rules of correct scientific method and administration. In terms of technological theory, standardized work processes and highly formalized organizational practices presuppose rather low levels of task uncertainty and high mutual dependence. For only if tasks are routine and work outcomes are predictable will formal rules and standard evaluation criteria be able to actually govern scientific and organizational practices. The orthodox philosophical model of rational science and neoclassical administration theory commit the same conceptual fallacy as the localist model of scientific production and the garbage-can model of organizing; the only difference being that the former pair describes the opposite extreme of "irrational" and idiosyncratic scientific and organizational practices. What the rational philosophical model of science and neoclassical theory describe is not the nature of good science or the one best way to organize; rather, it is scientific and organizational practices under conditions of low uncertainty and high dependence (e.g., Lawrence and Lorsch 1967).

We observe similar resemblances upon comparing the cognitive styles of epistemologies and organizational theories (see figure 4.3: "A Model of Epistemologies," p. 98). Neoclassical administration theory and realism share the same cognitive mentality or discursive style. Both realism and neoclassical theory are cognitive systems that emphasize uniformity and homogeneity more than diversity and pluralism. Neoclassical theory recommends universal principles of correct administration, and realism advocates universal rules of correct scientific method. Both cognitive systems have a strong normative and prescriptive, if not pontifical cognitive *habitus*, because they claim to be in possession of the only right ways to truth and scientific management.

Realism and neoclassical theory are less concerned with actual empirical variations than with methodological rigor and internal logical purity. Both are based on rigid and inflexible assumptions about how good science and good organizations should operate. Good organizations and good science follow standard and formal algorithms, regardless of differences in work processes, technologies, or environmental conditions.

We observe similar structural homologies between the cognitive styles of garbage-can/institutionalist models of organizing and the philosophy of the social sciences.[12] As opposed to realism and neoclassical theory, they both emphasize diversity and pluralism more than uniformity and standardization. Garbage-can theorists and most philosophers of the social sciences do not really believe that "scientific" administration and "scientific" social research are possible. Garbage-can models and the philosophy of the social sciences value methodological rigor and deductive consistency less than cognitive flexibility and openness to innovation. Both intellectual fields are less interested in standardization and routinization; rather, they suggest that methods of research and organizing be sensitive to variations in subject matters and issues. There is no one best way to organize, and no one best way to do research.

SUMMARY AND CONCLUSION

The technological approach to scientific organizations was developed in a critical review of the contingency tradition in organization studies. Several classical studies of organizational technology and structure were examined for their common conceptual denominators, which turned out to be the very variables of task uncertainty and mutual dependence used throughout the previous chapter. Task uncertainty and mutual dependence have similar effects on organizational structures and practices, regardless of whether scientific or nonscientific organizations are considered. High uncertainty and low dependence lead to more informal, disorderly, and idiosyncratic practices, and to more decentralized organizational control structures. Conversely, low uncertainty and high dependence lead to more formal, systematic, and standardized practices, and to more centralized organizational control structures. The technological tradition was synthesized into a comprehensive contingency model of organizations that will permit us to analyze the sciences as special cases of reputational work organizations.

The general theory of scientific organizations has the strong implication that the basic organizational dynamics do not differ between scientific and nonscientific organizations. The review of contemporary debates in organizational theory has shown that the current cleavages in organizational theory are conceptually very similar to those in the philosophy and sociology of science. Again, the technological model of scientific organizations was drawn upon to explain neoclassical administration theory and garbage-can models/institutionalism as describing opposite poles of what should also be treated as a *continuum* of structure and rationality. The broad applicability of the technological model demonstrates the comprehensive explanatory power of an organizational approach to science.

CHAPTER 6

Some Comparative Observations on Science and the Professions

In very general terms, scientific organizations belong to the "unit" production type described by Woodward (1965/1980). The structural features of such organizations are summarized in the top row of figure 5.4 in the previous chapter. The relevant *environments* of scientific organizations consist, of course, not only of "nature" or "reality," but also of other scientists and other scientific organizations. When generally compared with nonscientific organizations, these environments are rather heterogeneous and unstable (Lawrence and Lorsch 1967), since the reward system of science places a prominent emphasis on constant innovation. That is, the creation of novel knowledge, the emergence of new fields and specialties, the introduction of new research technologies, and changes in the interdisciplinary relationships between scientific fields generate rather unpredictable and unstable environments for scientists and scientific organizations. As a result, the overall level of task uncertainty is rather high in the sciences, the considerable variations between fields notwithstanding.[1]

As opposed to more bureaucratic forms of organization, scientists generally coordinate their work by mutual adjustments and consultation rather than by plans, schedules, or centralized command hierarchies. It must be kept in mind, however, that we would expect some scientific fields to resemble bureaucratic forms of organizing more closely than others so that there is no "one best way to organize" in science, either. Generally, however, scientific work is too uncertain and scientists' discretion over their work too high to coordinate and control research in a formal bureaucratic way. Not surprisingly, then, microsocial studies of laboratory life and controversies reveal constant informal negotiations between scientists over the meanings of data and the interpretation of theories.

Although research and lab directors do have some control over the setting of research goals and directives, individual scientists are largely autonomous in their day-to-day activities (Knorr-Cetina 1981; Lynch 1985).

Similarly, the communication of research outcomes and the certification of knowledge claims are largely a matter of negotiations among peers, not of authoritarian *fiats*. The decision to accept a colleague's work as the premise of one's own work, or, the decision to produce facts, is not a decision made by centralized command hierarchies, but by individual scientists in their collegiate networks. Of course, this is not to say that there is no stratification of authority in science, but the *coupling* of authority and decision making in science is not anywhere near as close and rigid as in more hierarchical types of bureaucratic organization.

In their analysis of the peer review system in the National Science Foundation (NSF), Cole, Rubin, and Cole (1978) tested the effects of scientists' location in the academic stratification system on the NSF ratings of grant proposals. Corroborating the argument presented above, the authors found that such indicators of scientists' academic status as reputation and visibility,[2] age, and the prestige of applicants' departments explained only little of the variance in NSF proposal ratings. Position in the academic stratification system also did not account for a large proportion of the variance in the actual NSF decisions to grant or deny funding to submitted research proposals. Cole, Rubin, and Cole (1978:122) conclude that the "merits" of scientific work are assessed largely independently from scientists' position in the stratification system. Merton and Zuckerman (1973) arrive at the same conclusion in their study of the effects of rank differences between scientists on the acceptance of manuscripts for publication in the *Physical Review*. That is, authority differentials undoubtedly exist in science, but their effects on peer inspection and review systems appear to be much smaller than in more bureaucratic and hierarchical types of organization.[3]

Scientific work, then, is controlled and coordinated more by negotiations among peers than by hierarchical supervision and influence differentials. As Collins and Restivo (1983b) point out, however, before the university organization of modern science stratification and coercive control played a much greater role. But the characteristic control form employed by modern scientific organizations is normative control; although, of course, material rewards are as essential in science as in other work organizations. But scientists are expected to perform at levels above and beyond what can be expected in exchange for material rewards alone. Scientists are expected to be driven by high intrinsic motivations, such as the "search for truth," and the "improvement of knowledge." The prominent emphasis on the "ethos of science" in Mertonian sociology of science reflects this strong reliance on scientists' identification with and commitment to their work. This is not to say that scientists actually follow the Mertonian ideals of communism, skepticism, disinterestedness, and universalism in their daily work (Mulkay 1979). In fact, scientists vehemently fight over priority claims, often adhere to "normal" research conventions rather dogmatically, invest their cultural and material capital in particular career strategies, and often follow local standards of research rather than the universal rules of scientific method. The ethos of science is probably hardly more than the scientific counterpart to the ideological rationalizations and ritual mysteries surrounding the other professions (Larson 1977). Sal Restivo (1988:208) calls this the "icons, myths, and ideologies" of modern science. "Normative control" is interpreted here in the more neutral sense of organizational control theory: science is a profession whose members are very dedicated to their work, and identify closely with the organization (R. Collins 1975).

The issue of normative control highlights some instructive differences and similarities between science and other professions. Generally, unlike bureaucratic organizations, the professions rely to a great extent on the normative identification of their members with their work. This strong normative orientation is instilled in members during extended periods of professional socialization, and is reflected in the professionals' view of their occupation as a "calling." Hence, all professions cultivate some regulative ethics or codes of honor (Moore 1970). These codes are expressed in the professional ideologies of altruism, universalism, and affective neutrality taken at face value in functionalist theories of the professions. As a result, lawyers appear to specialize in justice, doctors in health, artists in beauty, and scientists in truth. Professional ideologies justify members' practices, and they shield these practices from public inspection and criticism (Meyer and Rowan 1977). Elaborate rituals in the staging of professional work performances help materialize the ideological mysteries surrounding the professions. Physical props such as doctors' white coats, the secretive backstage consultations in courts, the artists' insistence on flashes of genius and inspiration required for aesthetic work, or the elaborate distinctions between "internal" and "external" areas of scientific research maintain auras of professional sacredness and nonmundane superiority.

The professions are the most ideological and ritualistic of all the occupations. This is so for two reasons. First, professionals generally enjoy a great deal of autonomy and discretion over their work. Unlike bureaucratic organizations, professional associations do not control the daily work of their members very closely; although, as will be seen shortly, there are considerable differences in the degrees of individual work autonomy between the professions. Also, the degree of autonomy seems to be lower for professionals practicing within larger organizations (Hall 1972). But generally, professionals are largely independent from direct supervision and order-giving hierarchies. Professional associations do have some control over practitioners by means of regulative codes, certification agencies, and credential institutions, but these controls affect general work standards and rights to practice more than everyday work.

In regard to the strong ideological and ritual character of the professions, high practitioner autonomy allows for a prominent *cultural* element in professional work (Greenwood 1972:12). While directly supervised and strictly controlled work is typically more "instrumental" and "technical" in character, high autonomy and discretion allow professional practitioners to cultivate elaborate symbols and rituals in the performance of work. Hence, a considerable part of professional work is devoted to the cultural staging and framing of that work. This cultural management of professional performance is probably most visible in the arts, but is present in all professions that involve frequent interactions between practitioners and customers or clients.

Also, the strong ideological and symbolic character of professional work is grounded in comparatively high levels of task uncertainty. All professions have to some extent established monopolies over their areas of expertise. Very powerful professions, such as science and medicine, control the markets for professional work and services by coupling the right or opportunity to practice to formal university training, credentials, and certification by professional agencies. Such monopolistic controls over areas of uncertainty are the basis for professional authority over the laity. This authority finds its most visible expression in the professional claim to possess some privileged and rational knowledge unavailable to the lay public. Professionals have, to varying degrees, acquired exclusive rights to define what is good and bad for a patient's health, what is in the best legal interests of a defendant, what may count as an authentic piece of art, and what knowledge corresponds to reality and deserves the title "scientific." High task uncertainty combined with monopolistic expertise account for the ideologically elaborated mysteries surrounding professional work. Since lay clients and customers are usually not competent to judge professional work, practitioners are in a good position to credibly claim that their services are in the best interests of the laity.

An important consequence is that the Mertonian celebration of the ethos of science results from the captivating and persuasive force of professional ideologies and mystery production. This celebration is part of the uncritical worship of the professions practiced by functionalism. All professions, not just science, use ritual mysteries and ideological idealizations to shield their practices from critical public inspection. This is the important lesson to be learned from institutionalist theories of organizational structure. Normative ideologies are an integral part of the professional self-presentations of science, but to start a sociology of science with buying into its ideological mystifications is hardly a reasonable strategy.[4]

Despite these common properties shared by all professions— ritual symbolisms, practitioner autonomy, and monopolistic expertise—there exist some crucial differences between them. I shall now turn to these differences, and focus on variations in the familiar variables of task uncertainty and mutual dependence.

TASK UNCERTAINTY AND STRATIFICATION

Although the general level of task uncertainty is higher in the professions than in more bureaucratic organizations (Scott 1981:241), professions differ in the extent to which work is routinized. Task uncertainty appears to be higher in science and the arts than in the legal and medical professions, since the former place special emphases on innovation and creativity. Scientists and artists acquire the highest reputations for novel contributions, although the need to coordinate one's work with that of colleagues—and the dogmatic force of scientific and aesthetic conventions—temper innovativeness and creativity. Radical innovations occur only rarely precisely because they break with established traditions and, hence, cannot easily provide a resource for ongoing collegiate work. But scientists and artists are expected to advance knowledge and create new forms of aesthetic expression, thereby increasing the overall level of task uncertainty in these professions.

Doctors and lawyers, on the other hand, generally deal with more predictable problems. Once their formal training is completed, they face fewer pressures to constantly update their knowledge and practical skills. Of course, doctors and lawyers must to some extent respond to advances in medical knowledge, or to changes in legislation and jurisdiction. But neither the rate of change nor the pressures to innovate one's practices accordingly appear to be as high here as in science and art. The day-to-day work routines of doctors and lawyers require no *systematic* connections to *permanent* advances in medical and legal knowledge. In the case of lawyers, changes in legislation and jurisdiction can only be made and implemented by large and inert bureaucracies, which greatly reduces the speed at which such changes may affect lawyers' daily practices.

In science and art, on the other hand, practices and technologies become outmoded rather quickly, either in response to frequent innovations produced at "research fronts" or to rapid fluctuations in what is considered "fashionable" in art worlds (Becker 1982:300ff.). That is, scientific and artistic communication systems diffuse innovations more rapidly to individual practitioners. Gaining and maintaining high reputations requires scientists and artists to constantly keep up with current advances and innovations in their fields. Doing "state-of-the-art" work is a key requirement for publishing in the most prestigious scientific journals and for remaining *en vogue* in the relevant artistic communities. As a result, scientific and artistic work generally seem less certain and routinized than the work of doctors and lawyers.

There are, however, differences in the degrees of task uncertainty faced by various segments of practitioners *within* a profession. These differences appear to correlate with the internal stratification of various professional groupings. That is, the core areas with the highest task uncertainty seem to be controlled by the most prestigious elite groups within a profession. The higher the task uncertainty faced by a professional grouping, the more prestigious its status within the larger community.

In science, for example, those scientists working in what Price (1986) has called the "research fronts" enjoy higher prestige and reputation than those working on more routine and peripheral problems. Scientists working in the rapidly advancing core areas of a discipline not only control areas of high task uncertainty, they also define where the field as a whole is moving since research front work is most uncertain, it will also be the least bureaucratic and methodological. Hence, these scientists enjoy the highest prestige and influence, and they obtain the most visible awards:

> Definitions of what constitutes a prestigious specialty appear to be a product of the assessment by the field as to the areas in which significant problems exist as well as the extent to which solutions of the problems are difficult, that is, require relatively scarce talent to solve. (Cole and Cole 1973:44)

Similar relationships between task uncertainty and stratification seem to hold for the other professions considered here. Howard Becker (1982:228–46), for example, has shown that a crucial social distinction in the art world is that between "integrated professionals" and "mavericks." Integrated professionals are those artists who have close ties to established institutions of artistic production and distribution. They produce highly conventional works of art by means of routine technologies and traditional aesthetic devices. Integrated professionals rarely produce novel art, for just like "normal" scientists, they "operate within a shared tradition of problems and solutions (Becker 1982:230)." Integrated professionals represent the great majority of artists who do not attract a lot of attention by spectacular revisions in accepted aesthetic practices. They are commonly regarded as being interchangeable and downgraded as competent but uninspired. Integrated professionals, then, are the unpretentious "technicians" of art worlds who produce the vast majority of all works of art, but whose work is not very visible and prestigious. However, the routine work of integrated professionals keeps the mundane *organizations* of art worlds busy: their art fills the galleries, occupies the critics, and sells to the audience.

Integrated professionals, then, produce conventional art in a very routinized way. Like scientists working on more standard problems with rather predictable outcomes, integrated professionals do not enjoy high reputations, visibility, and prestige. According to Becker (1982:233ff.), mavericks, on the other hand, are artists who produce novel art. Mavericks find the conventions of aesthetic production constraining and, hence, work outside the institutions of mainstream art. Mavericks often form separate and esoteric social networks, independent from the worlds of integrated professionals. They sometimes even create new organizations of aesthetic production and distribution, for their work typically does not fit into the dominant modes of producing and disseminating art. Although many mavericks are not taken seriously, or hardly recognized at all, some acquire high reputations and prestige for creating novel forms of aesthetic expression. For the innovative work of mavericks corresponds more closely to the widespread expectation of artists' work bearing the signs of creative genius than does the routine art of integrated professionals.

That is, just like in science, highly innovative and uncertain artistic work gains the highest visibility and prestige. However, just like novel science, innovative aesthetic work must to an extent be understandable in terms of accepted traditions and conventions. Work that departs from tradition too radically will not be taken seriously, or will not even be recognized as "competent" science and art. In science and art innovation must be tempered by tradition. Kuhn (1977) has called this precarious balance "the essential tension."

Generally, then, scientific and artistic elites control the most uncertain and innovative areas of scientific and artistic production. Stratification correlates with task uncertainty and innovativeness. To an extent, this also appears to be true in the medical profession, for the most prestigious medical work is performed in the most complex and uncertain core areas of the discipline, such as brain surgery, organ transplants, or elite psychiatric counseling. Generally, hospital-based medical specialists outrank the traditional generalist with a solo practice ("family doctor"), especially when the former have some ties to medical research centers and university departments (Larson 1977:164f.). The medical elites monopolize the highly specialized and uncertain core areas of the discipline and leave the great bulk of more routine cases to less specialized and prestigious general practitioners. Hence, the medical referral system usually channels patients from less prestigious general doctors to medical specialists. The division of labor in hospitals also follows this pattern: the highest ranked specialists perform the most complex and rare tasks, while the majority of doctors work on more frequent and predictable routine cases.

In the legal profession, however, a different system of stratification seems to operate. In their study of Chicago lawyers, Heinz and Laumann (1982) found that the most important and visible status cleavage in the bar separates lawyers representing individual clients from law firms working for large and powerful corporations. The legal elite largely consists of corporate lawyers, which Heinz and Laumann (1982:331) interpret to mean that the prestige of lawyers *derives* from the wealth and power of their clients. The more wealthy and powerful the clients, the more prestigious their legal consultants, and organizations are generally more wealthy and powerful than individuals. That is, unlike in science, art, and medicine, the status stratification in the legal profession corresponds much more closely to the general wealth and power differentials found in the larger society. This is why Larson (1977:166) calls the legal profession the "epitome" of social stratification.

It is thus useful to distinguish between "internal" and "external" controls of a profession's stratification system. This distinction corresponds to Johnson's (1972) distinction between "collegiate" and "patronage" occupations.[5] By "internal" I mean that professional communities themselves, not the statuses of clients or customers, determine a practitioner's position in the professional stratification system. Internal control seems to be present, to varying degrees, in science, art, and medicine. For, in these professional fields, it is the complexity, novelty, and uncertainty of the problems worked on that determine practitioners' ranks in the prestige hierarchy. The legal profession, however, is divided along the "external" lines of power and wealth differentials between various client groups. In Johnson's (1972) terms, science, art, and medicine are collegiate professions, whereas the law, together with fields such as architecture and engineering, is a patronage profession.

This explains why many scholars have observed that the legal profession is not very coherent and powerful *as a profession*, al-though, of course, individual lawyers and law firms have consider-able influence over their clients and over the legal process in gener-al (Larson 1977; Heinz and Laumann 1982; Rueschemeyer 1972). But in terms of the classical properties of powerful professions—self-governance and autonomy, collegiate control over practi-tioners and customers or clients, unified cognitive system, and disciplinary monopoly—the legal profession is less integrated and coherent than, say, science or medicine.[6]

Consequently, Heinz and Laumann (1982:323) observe the paradox that the most powerful corporate lawyers have the least autonomy and control over their work, while lawyers representing individual clients come a little closer to the ideals of independent professionalism. Corporate lawyers work for powerful organiza-tions that closely control and supervise legal work, often through corporate house counsel. Lawyers serving individual clients, how-ever, are more independent because individuals usually have less time, knowledge, and competence to control their legal affairs. This situation is paradoxical because the less prestigious individual-client lawyers have *more* work autonomy and discretion than their more prestigious corporate colleagues. The reason for this paradox is, I suggest, the *external* control of the status system in the legal profession. Since lawyers' status is *derived* from client status, the allocation of prestige does not follow the more *internal* criterion of whether the legal work done is more routine, predict-able, and certain, or more complex, innovative, and uncertain. *As a profession*, lawyers are not a very powerful group because of this external control.

Of course, the other professions considered here are not en-tirely independent from external controls of their status and pres-tige systems. It remains true that *all* professions display a strong upper-middle class bias, that the most renowned artists do not produce for mass markets, and that the most prestigious doctors are unlikely to work in poor-neighborhood community clinics. But the status of scientists, doctors, and artists is not so much derived from the status of their clients, patients, or customers. That is, the stratification system in these professions is based more on internal or collegiate mechanisms for allocating prestige and reputation, such as working on highly uncertain, complex, and innovative problems. The most prestigious scientists do not necessarily come from the wealthiest families,[7] some artists may even lose prestige if their work is "too commercial," and the reputation of the highly specialized neurosurgeon does not suffer from his or her patients being poor and powerless.

There is, however, one divergent interpretation of the relationships between prestige, task uncertainty, and client status that would make the differences between internally and externally controlled professions less drastic. Let us reconsider the case of law. Generally, lawyers representing individual clients deal with more repetitive and routine problems, such as divorce and personal injury cases. Corporate lawyers, on the other hand, work on the more esoteric and complex legal problems only large organizations can have. These problems are more unique and typically require more extensive legal knowledge and skills than those which individual clients present. That is, the prestige differentials between lawyers representing individual clients and corporate lawyers representing large organizations might not so much reflect the wealth and power differences between these two client groups, but rather the distinction between routine and nonroutine tasks:

> The wealth and intellectual challenge variables are, thus, closely interrelated, and it is therefore difficult to determine whether the prestige of the fields is to be attributed more to client type or to the differential opportunities that the fields present for the exercise of intellectual skills. (Heinz and Laumann 1982:331)

In this interpretation, the stratification system in the legal profession resembles that in the other professional fields: status cleavages separate practitioners working on more predictable and routine problems involving low task uncertainty (normal scientists, integrated professionals, medical generalists, individual client lawyers) from practitioners dealing with more innovative and complex problems involving high task uncertainty (research front scientists, mavericks, medical specialists, corporate lawyers). In any case, external or nonprofessional controls of the status system are no doubt stronger in the legal profession, and Heinz and Laumann's (1982) empirical tests also suggest that the client status factor is more important for determining lawyers' prestige than the task

Professional Status Derived from

	Internal Criteria: Collegiate Professions	External Criteria: Patronage Professions
Low	Integrated Professional Artists Normal Scientists Medical Generalists	Individual Clients Lawyers
High	Maverick Artists Research Front Scientists Medical Specialists	Corporate Lawyers

Prestige (row axis label between Low and High)

FIGURE 6.1   Status Differentials in Four Professional Fields

uncertainty factor. Figure 6.1 summarizes our discussion of status differentials in four professional fields.

MUTUAL DEPENDENCE AND PROFESSIONAL WORKSTYLES

Professional fields not only differ in the degrees of task uncertainty invested in their work, they also establish varying degrees of mutual dependence between practitioners. In the present context, "mutual dependence" comprises two related but distinct variables. *Entry restrictions* indicate the extent to which corporate institutions control the markets for professional services. Most importantly, professional bodies control rights to practice and allocate work opportunities to practitioners. Strong professions monopolize rights to practice by means of licensing and status certification boards so that all "legitimate" practitioners must be formally acknowledged by some professional agency. Sometimes, this professional authority is backed up by the coercive force of the state. In particular, licensing boards and certification agencies control access to jobs, work facilities, educational credentials, and customers or clients. Also, strong professions, such as medicine, regulate the number of "legitimate" practitioners, and hence control the amount of competition and the prices for professional services.

Generally, the more severe the entry restrictions, the more exclusive, elitist, and powerful the profession. Severe entry restrictions monopolize access to professional work opportunities and force the laity into rigid dependence relationships that are the source of professional power. The fewer alternative options the laity has to obtain professional services, the more powerful are the professional groups who control service markets. Entry restrictions can be very tight, such as in medicine and modern science, or they can be rather loose, as in most contemporary art fields. In the former case, services are offered on the seller's market, while most art is sold and bought on the buyer's market.

Severe entry restrictions unify professional groupings. Due to tight professional control over rights to practice and market access, powerful professions are internally more homogeneous than less powerful professions. Powerful professions can impose more uniform educational standards upon the curricula, they can set common rules of market performance, and they can establish shared cultural rituals of professional self-presentation. By controlling access to service markets, powerful professions can deny "illegitimate" professionals and competing groups rights to practice. In this way, entry restrictions also affect the second aspect of mutual dependence considered here, "social density."

While entry restrictions determine the general market and competitive structure of professional fields, "social density" indicates the degree of collegiate control over the everyday work of individual practitioners. In the professions, collegiate control is generally tempered by the high work autonomy and discretion granted to individual practitioners, but the amount of control varies between professional fields.

The neo-Durkheimians have shown that the higher the social density found in groups—or, in the present context, the tighter the collegiate control over practitioners—the more collectivistic and unified the shared cognitive structures and orientations.[8] Tight collegiate control leads to a strong emphasis on "objective" and "correct" ways of practicing one's profession. Close collegiate inspection and supervision will temper individual innovativeness and creativity, for the group watches closely over an individual's performance. There are rather rigid and inflexible rules for "proper" performance as defined by the group, and there is not much tolerance for deviance and dissidence. The group has a great deal of authority over individual performers and strongly insists on the essential difference between right and wrong. Tight collegiate control thus reduces individual autonomy and discretion. Individual contributions must observe collective standards and fit into shared collegiate practices.

When social density and collegiate control are weak, on the other hand, individual creativity and innovativeness are more highly valued and rewarded. There exist only loose and ambiguous standards for doing one's work, and these are subject to individual definition and interpretation. As a result, distinct local and individual styles of work may develop in loosely structured professions. The group is not strong and cohesive enough to enforce conformity, or even to decide what "conformity" to the group norms would actually imply. Professions with weak collegiate control structures are rather pluralistic, cosmopolitan, and individualistic, and their workstyles are more informal, discursive, and controversial. There is a great deal of conflict and debate over fundamental or philosophical issues, such as over the nature of social action in theoretical sociology, or over the relativism of aesthetic judgments in art. Tightly integrated professions, on the other hand, worry less about foundational issues and are more pragmatically concerned with piecemeal improvements in work technologies, and with slow but steady advances in objective knowledge.

Combining the two dimensions of mutual dependence—entry restrictions and social density or collegiate control—yields the social map of professional fields and subfields depicted in figure 6.2. The concrete location of particular professions in this map is rather tentative, but the important point here is conceptual: differences in the work routines and cognitive practices between the professions do not so much stem from ontological differences in subject matters or areas of expertise, but rather from differences in social organization.[9]

From figure 6.2 it appears that an important factor explaining variations in both density and entry restrictions is the level of concentration in the resources for professional production, such as reputation, lab equipment, medical equipment, and publication

	Entry Restrictions	
Resource Concentration	High	Low

High	Research Front Science	
	Medical Specialists (Hospitals)	
		Movies
	Corporate Lawyers	
Social Density/ Collegiate Control		
		Social Theory
	Medical Generalists (Private Practice)	
Low	Private Lawyers (Individual Clients)	
		Literature
		Poetry

FIGURE 6.2   Collegiate Control in Professional Fields

space. If the symbolic and material resources for professional production are highly concentrated, then practitioners depend very
closely on each other, and especially on the professional elites who
control these resources (Traweek 1988).[10] That is, a high level of
resource concentration will tighten entry restrictions and increase
social density or collegiate control at the same time. High resource
concentration strengthens a profession's monopoly over its area of
expertise and, simultaneously, reduces the extent to which individual practitioners are independent from collegiate control. If, on the
other hand, resources are more widely dispersed and decentralized, then access to work opportunities will be more readily
available, and practitioners will enjoy more individual work autonomy. Low resource concentration permits competing professional
and nonprofessional groupings to practice independently and thus
weakens the networks of collegiate control over practitioners.

It is important to notice at this point that differences in resource concentration, density, and entry restrictions *cut across* distinctions between various professional fields so that we would expect, for example, some scientific fields to resemble particular
areas in medicine *more closely* than they resemble other scientific
fields, some branches of art to resemble science more closely than
they resemble other arts, and so forth.

That is, the important *sociological* differences between professional fields cut across the traditional categories for identifying professions according to tasks, clients, or areas of expertise. From our organizational perspective, the most important differences are not those between doctors, lawyers, artists, and scientists, but between professionals working in fields with severe entry restrictions, high resource concentration, and dense social networks versus professionals working in fields with lax entry restrictions, widely dispersed resources, and weak collegiate controls.[11] For these structural variables determine not only a profession's general market and competitive status, they also affect practitioners' everyday work experiences, such as the amount of work autonomy granted to individuals, the degree to which tasks require practitioners to regularly interact and cooperate with colleagues, and the extent to which professional colleagues control career opportunities.

Consider again figure 6.2. From our perspective, the crucial dividing line separates professional fields with high scores on entry restrictions, resource concentration, and density from fields with low scores on these variables. Take, for example, hospital-based medical specialists versus general doctors with solo practices. Although *formal* entry restrictions are probably equally severe for both groups,[12] their everyday working environments differ quite considerably. Due to a high level of resource concentration, hospital-based medical specialists depend more closely on colleagues in their everyday work. Hospitals, not their doctors, own the medical equipment. While general solo practitioners are largely autonomous and independent from collegiate inspection and evaluation in their daily work, hospital work involves a great deal of cooperation and mutual adjustments between various specialists. Collegiate control is especially tight in medical situations involving the actual copresence of several specialists, as in surgery, but is also considerable in the processing of patients through various interdependent specialist divisions and care units. Whereas general practitioners work on the "whole patient," and hence may still cultivate an ideology of altruistic dedication to the patient's general welfare, specialists in hospitals work on particular parts of the sick body, and hence must coordinate their activities with other specialists. General practitioners have patients, specialist doctors have cases.[13]

Keeping in line with neo-Durkheimian reasoning, we would expect hospital-based specialist communities to have quite different cultural outlooks and cognitive orientations. In particular, medical specialist communities are more likely to follow recent developments in medical research. Specialist elites frequently have close ties to research communities and sometimes are involved in research and teaching themselves. Due to the close coupling of specialist activities in the hospital, changes in medical practice will, much like normal accidents (Perrow 1984), spread quickly throughout specialist networks. Hence, medical specialists will tend to emphasize the "scientific" nature of their work and will be very confident in the superior status of "scientific" medicine. Since specialist communities are rather dense and cohesive, they will tend to develop a strong *esprit de corps*. Because specialist doctors are also more prestigious than medical generalists, the group will cultivate a fairly elitist and self-confident caste consciousness. Since they work only on particular parts of the body, and since they interact with other specialists more closely than with patients, specialist doctors will tend to define themselves as members of elite collegiate groups, not so much as servants to the public.

General practitioners, on the other hand, are much less homogeneous and cohesive as a group, for their work patterns do not require a great deal of collegiate cooperation and interaction. Hence, generalists are not a very strong and well organized group within the medical profession (Larson 1977). Since they are rather independent from direct supervision and control, general doctors will be more likely to develop more idiosyncratic and personalized styles of work, much like scientists in loosely coupled fields. This is why patients often feel uneasy about changing their general doctors. General doctors interact more frequently with patients than with colleagues, and hence their *conscience collective* will be somewhat less remote from the laity. Also, generalists are not as likely as their specialist colleagues to closely follow and implement advances in medical research, and they have weaker ties to scientific communities. By definition, general doctors do not specialize in particular diseases or parts of the body, and hence advances in medical research, which tend to occur in highly specialized fields, will not be as relevant for the generalist as they are for the specialist. As a consequence, general doctors will rely to a greater extent on accumulated practical experiences and informal personal knowledge than on objective scientific expertise.

In sum, medical specialists and generalists inhabit quite distinct professional worlds. These worlds differ in the ways practitioners do their work and relate to their colleagues and patients. I believe a similar argument can be made for the legal profession: the social cleavages between individual client lawyers (who are comparable to medical generalists) and corporate lawyers (who are comparable to medical specialists), run so deep that Heinz and Laumann (1982) speak of different "hemispheres." However, I shall not pursue this issue further here, but return instead to science and compare its structural characteristics to the profession of art.

This comparison is especially instructive given that science and art have traditionally been opposed to each other for epistemological reasons. In this view, scientific knowledge follows the objective rules of scientific method, corresponds to reality, cumulates toward Truth, and is subject to the impartial and universal forces of Reason and Reality. Aesthetic knowledge, on the other hand, is a matter of indisputable tastes, changes according to fashions, is borne out of creative imagination, and can only be evaluated by arbitrary judgments. From an organizational perspective, however, the differences between science and art emerge from differences in the social organization of scientific and aesthetic production. As we shall see, some sciences look more like arts, and some art forms are rather scientific.

SCIENCE AND ART: AN ORGANIZATIONAL COMPARISON

In terms of the three variables of resource concentration, entry restrictions, and social density, science is a very strong collegiate profession. In most modern sciences, access to research facilities and specialty networks requires extended professional socialization, educational credentials, and some status position within academic labor markets. Ever since the concentration of research in the reformed Prussian university system, it has become increasingly difficult for lay practitioners and amateur scientists to participate in "serious" research. Nowadays it is virtually impossible for nonprofessional practitioners to contribute to scientific research, at least in most natural sciences.[14] Research contributions require access to research facilities and specialty networks, and amateurs without appropriate university training, credentials, and labor market status find themselves largely excluded from these resources. The process of professionalization has ended the gentleman-amateur tradition of wealthy private individuals conducting experiments in the socially isolated space of home laboratories. Unlike access to medical markets, entry restrictions to modern science are not backed up by state sanctions against "illegitimate" practitioners, but in effect, amateurs are excluded from "established" science.[15]

Not only are entry restrictions severe in modern science, there is also a fair amount of social density or collegiate control over practitioners. Science differs from all other professions in that reputations, access to the means of scientific production, and status positions very closely depend on collegiate evaluations of one's work. In science, reputations are earned by publishing "significant" and "innovative" contributions to research, and decisions to publish are made by professional colleagues. Science is unique in that peer review and inspection systems are absolutely crucial in determining practitioners' reputations, visibility, and work opportunities. Collegiate control is generally high in science because the outcomes of everyday work cannot earn reputations and academic status without ongoing collegiate inspection and approval.

To be sure, collegiate control is essential in most professions because licensing and certification agencies grant rights to practice and access to service markets. But once a doctor, or at least a general doctor, or a lawyer are accepted as "legitimate" practitioners, their everyday work is less dependent on ongoing collegiate inspection and approval. The outcomes of scientific research are subject to *persistent* peer approval, and often peers interfere even more directly with one's work, as is the case when reviewers make concrete suggestions for the revision of a conditionally accepted scientific paper.

I have argued above, however, that the degree of social density or collegiate control varies even within professional fields. Hospital-based medical specialists are more subject to collegiate inspection than generalist doctors with solo practices. Similarly, corporate lawyers working in large law firms must probably coordinate their work activities with one another more closely than lawyers serving individual clients. Generally, whenever tasks require the actual copresence of various professional specialists, or whenever these specialists must routinely coordinate their work, then density, and hence cognitive uniformity, will be higher.

In science, experimental fields performing a great deal of laboratory work generally involve more copresence and work coordination than do nonexperimental fields. Nonexperimental and theoretical work is more lonely mental production. Laboratories, however, are sites of constant interaction and negotiation between scientists who gather around the workbench and use the technical equipment together. This helps explain why labs often develop highly localistic cultures of research (Knorr-Cetina 1981; Latour and Woolgar 1986). Extended periods of copresence and interaction between scientists are likely to generate particular and group-specific workstyles and techniques. Copresence always involves the duality of perception and communication typical of interaction systems (Luhmann 1984; Fuchs 1989), and this duality explains why members of local groups share practices with one another, but not as much with members of other groups.

Science, then, is a very strong collegiate profession. By controlling publication decisions, peer gatekeepers not only determine the fate of a completed scientific paper or research proposal, they also influence every single stage of a research program. For scientists will select problems, literatures, theoretical approaches, research techniques, and presentational devices *in the light of* expected publication and funding chances. Scientists will not organize their research in a way that reduces these expected chances, and hence, they adjust their entire work to anticipated collegiate responses.

That is, *other scientists* decide whether or not to accept a colleague's work for publication and funding. Only *other scientists* can decide whether or not to incorporate one's work into their own research, and thus only *other scientists* can turn each other's statements into black boxes or facts.[16] In this limited sense, science is a profession without customers or clients.[17] In other professions, customers and clients have greater control over professional work. The powerful organizational clients of corporate lawyers will make very sure that the legal work being done serves their best interests, and doctors' patients generally know when they are feeling better. That is, in the legal and medical professions the customers of services limit the extent to which professional control may proceed in a purely self-referential way. Collegiate control in science, on the other hand, is generally less influenced by lay concerns.

Some scientific fields, however, are more professionalized and self-referential than others. I would suggest again that entry restrictions and collegiate control are most severe in fields where the symbolic and material means for scientific production are most concentrated. Resource concentration tends to be especially high in fields using very expensive equipment (Traweek 1988). The more expensive and scarce the research equipment, the more concentrated the resources for scientific production, and the tighter the collegiate control over work. Such densely organized areas constitute what Price (1986) has called "research fronts." Research fronts or "invisible colleges" (Crane 1972) are small specialty networks located at a few elite departments and laboratories. Scientists in these highly prestigious and exclusive networks work on innovative problems in the core areas of a field and, as a consequence, control where the field as a whole is moving. Scientists at research fronts proportionately publish the most papers in a field, and their work is highly visible and frequently cited (Cole and Cole 1973; Zuckerman 1977). Research front communities constitute the top stratum of the scientific stratification system and, hence, informal entry restrictions are rather severe. Due to their small size and elite status, access to research fronts is very competitive and limited.

As a result, these networks have a great deal of control over individual practitioners and often recruit their members through master-apprentice ties (Zuckerman 1977:115). Group pressures for scientific conformity will be rather intense, although, of course, the high task uncertainty in areas of innovation encourages some degree of individual creativity and autonomy.

Due to high collective cohesion and density, research front groups will have a great deal of authority over individual scientists. Intense collegiate control will produce high group consciousness and confidence in the collective advance of knowledge. The tighter the intersubjectivity between group members, the stronger is the emphasis on the objectivity of methods and knowledge. Research front scientists exchange a lot of preprints, and their published works are highly interactive (Price 1986). Since research front scientists constitute the scientific elite, they will identify closely with the official norms and ideals of science as an institution. This celebratory attitude toward the merits of science as an institution is, of course, most obvious in the award lectures of Nobel laureates (Mulkay 1985).

Price (1986) and others (R. Collins 1975; Whitley 1984; Hargens 1988) have argued that closely integrated collegiate networks working at rapidly advancing research fronts are more typical of the natural—especially the physical—sciences, and less so in the social sciences and humanities. In these latter fields, entry restrictions are less severe, and the resources for intellectual production are less concentrated. Particularly in the nonexperimental social sciences and humanities, the material means for intellectual production are much less expensive and are more widely distributed across communities. With the possible exceptions of small group experimental research and large-scale survey projects, work requires less actual copresence of several practitioners. Teamwork among copresent scientists does occur more frequently in experimental group research and massive survey projects, especially in those relying extensively on the collection and computerized processing of quantitative data, but is certainly less frequent and widespread than laboratory work in the natural sciences.[18]

As a result, collegiate control in the social sciences and humanities is generally more decentralized and weaker than in the natural sciences. Although peer review and inspection systems remain very important for allocating reputations and work opportunities, the more decentralized control structure favors the coexistence of multiple paradigmatic schools and approaches, as is the case in contemporary sociology (Mullins 1973; Ritzer 1980). Collegiate control is fragmented there into a large variety of specialty and regional journals, associations, and scholarly networks (Turner and Turner 1990). Reputations and work opportunities can be gained from a large variety of organizational sources, and so the overall paradigmatic cohesiveness in the social sciences and humanities is rather low (Hargens 1988). Whitley (1984:158ff.) calls such fields "fragmented adhocracies." In the third chapter, I have argued that such fields tend to produce rather discursive, philosophical, and historical styles of intellectual work. Since mutual dependence is rather low, decentralized fields with multiple coexisting reputational organizations engage in conversation rather than in fact production.

We are now in a good position to compare the organizational control structures and corresponding intellectual styles of science to art. The soft sciences and art receive similar scores on the three variables structuring our comparative social map of the professions; that is, entry restrictions, resource concentration, and social density or collegiate control (see figure 6.2, p. 157). Correspondingly, I shall argue that soft sciences and art engage in similar styles of intellectual work. This similarity has been observed, albeit for the wrong reasons, in the currently fashionable postmodern equation of sociology with textual narrative and rhetoric. Since the postmodern equation rests on the textuality of science, I shall focus this part of my discussion on literature.

ORGANIZATIONAL CONTROL IN MODERN LITERATURE

Modern literature scores low on each variable measuring the degree of organizational and conceptual integration in professional fields (see figure 6.2, p. 157). To begin with, as a profession literature has no formal entry requirements so that no one can be legally prohibited from becoming a writer. Writing does not require standardized professional socialization, or educational training and credentials. Of course, this is not to say that anyone has an equal chance of becoming a successful writer. As is true for most professions, practicing professional writers usually have upper middle-class backgrounds, which sometimes assures them of some independent wealth and gives them the complex linguistic skills required for practicing literature as a profession. In addition, some ties to established literary circles will greatly facilitate one's career as a writer. But in comparison to medicine or science, literature is rather open to amateurs.[19]

Literature is a very noninstitutional profession. There exists no genuine labor market for writers, and most writers are not employed by organizations.[20] The extreme case is poetry, which is often reproduced and distributed by the poets themselves (Becker 1982). Literature is an extremely "private" profession, which is why writers' guilds and literary associations are difficult to organize and often ridden with internal ideological conflict. Consumers do not consider the services of writers as indispensable as medical or legal services because reading is a leisure activity. Hence, writers have a very weak market position and hardly any political bargaining power as a group.[21]

This noninstitutional and private character of the literary profession is to a large extent due to its very low level of resource concentration. Unlike science, literary production does not require expensive equipment or the copresence of several practitioners. Literary work is socially isolated and lonely mental production.[22] Literary styles are thus often highly idiosyncratic and individualistic, as is indicated by the virtual absence of coauthorships in literature. The individual author is the point of reference in the interpretation of his or her work. Literary critics talk about the "late Hemingway," the "mature Kafka," and the "Hesse after his Indian journey." Critics and interpreters link stages in writers' works to biographical experiences.[23] Unlike in science, there exists no clear separation between writers' personalities and their work. Pictures of authors can often be found on the jackets of their books. Literary production is not even expected to follow some "objective rules for aesthetic production," for literary insight is said to be borne out of poetic imagination and creative intuition. The public has a great interest in the personal lives of famous writers and closely follows their personal tragedies and excessive lifestyles. Whereas scientists step back from their intellectual products and let the facts and other scientists speak for themselves, writers' personalities permeate their work. Very famous writers are surrounded by personality cults that celebrate the heroic individualism of a unique but lonely and often alienated mind. Writers are expected to be extraordinary individuals, and sometimes they are even expected to be estranged from the mundane world.[24]

The highly individualistic and discursive style of literary production is also due to a very low level of collegiate control. There exist no formal systems of peer review and inspection in literature. In literature, the closest counterpart to scientific peer review is literary criticism, but the crucial difference between the two is that literary critics do not directly control publication decisions. Literary critics might influence a book's commercial success and, in this way, affect literary production indirectly via the interests of publishers to sell. But literary criticism occurs only *after* a work of literature has been published and appears less influential in shaping every stage in its actual production. Also, literary criticism often adopts the highly individualistic style of literary production itself (Hauser 1982:472): the standards for separating good from bad literature reflect personal tastes and preferences more than objective aesthetic knowledge.

But most importantly, literary critics differ from scientific peers in that their evaluations are not primarily addressed to *other* literary critics or to the writers themselves, but to the general audience.[25] And the modern mass markets for literature are amorphous enough to permit even esoteric work to find its niche and a "faithful" audience. Hence, evaluative standards will be much more ambiguous and loose in literature than in science, where criticism is made by and directed toward one's professional colleagues. In science, the producers, critics, and consumers of knowledge are members of the same general collegiate network, which greatly unifies the cognitive standards of scientific work. In literature, on the other hand, producers, critics, and consumers all form distinct social groupings with different preferences and loyalties. As a result, the standards of literary production tend to be more diverse, ambivalent, and controversial.

Collegiate control in literature is very weak, then. As a consequence, unlike scientists writers do not routinely use each other's insights as resources for their own creative work. Writers do not reference the work of other writers in their own writings, and they do not routinely use each other's statements as black boxes on which to build their own statements. Unlike scientists, writers cannot mobilize a whole array of supporting textual and nontextual agents to raise the costs for objecting to their statements.[26] Hence, literature does not produce facts, it produces "conversations" between highly personal and often incompatible aesthetic experiences. For organizational rather than epistemological reasons, then, literature is essentially noncumulative. This is not to say that there are no discernible traditions and conventions in literature that would identify shared literary styles, such as realism, naturalism, or modernism. But the control structures of literary production do not require the routine mutual adjustments and coordination between statements so typical of modern science.

In sum, literature is a very weak collegiate profession. There exist no formal entry restrictions, and the material means of literary production are widely dispersed. Collegiate control over practitioners is extremely loose, for literature lacks formal peer review systems controlling publication chances. Literary criticism forms a separate organizational network, for aesthetic critiques of literary works are primarily addressed to the general lay audience, not to other critics or the writers themselves. Writers do not routinely adjust their own production to that of colleagues, and they are much less likely than scientists to pay a great deal of attention to critics. In addition, the amorphous mass market structure of modern literature holds niches for a large variety of literary styles. As a result of these organizational conditions, modern literature is lonely, highly personalized, individualistic, and discursive mental production.

THE POSTMODERN EQUATION OF SCIENCE AND LITERATURE

Given the organizational and conceptual structures of modern literature, it should not come as a surprise that the postmodern turn in certain branches of current social theory has likened science and sociology to literature (e.g., Agger 1989; Brown 1987; Edmondson 1984; Green 1988).[27] The postmodern view focuses on the textuality shared by science and literature and regards common textuality as the reason for likening science to literature. As texts, science and literature use certain literary methods and stylistic devices to present coherent images of reality and to persuade the audience of their authenticity. In this view, the rhetoric of science and sociology is not just a neutral medium for communicating true ideas of reality (Gusfield 1976). Rather, postmodernists claim that textuality *creates* reality by means of rhetorical persuasion. There is no independent reality "out there" that could simply be mirrored by valid scientific descriptions, for objective reality is itself a textually created fiction.

Scientific texts create the fiction of reality and of corresponding to something "out there" by fostering the illusion that facts speak for themselves. Authors withdraw from the dramatic frontstages of actual scientific construction, and hide behind the objective methods that appear to do nothing but help reality describe itself. Postmodern social theory thus turns into literary criticism: it inspects the stylistic ways in which sociological and scientific texts use language to create persuasive images of reality, and to couch the ideological beliefs of their authors in neutral terms of fact (see Brown 1987; Edmondson 1984; Green 1988).

Postmodern approaches in social theory usually draw upon the "postpositivist" turn in the philosophy of science. Postpositivism criticizes the metanarrative of scientific rationality providing the objective foundations for occidental culture. Postpositivism denies that scientific knowledge has any privileged access to objective reality and that science follows some superior rational logic of research. Just like mundane knowledge, science is seen as a social construction and institution. Scientific knowledge is socially constructed by scientists working at particular places at particular times. Science is a social event. Hence, postpositivist philosophy of science no longer concerns itself with the epistemic conditions under which rational science will correspond to external reality and approximate truth. Rather, the attention shifts to science as an ordinary narrative and mundane social institution.

It seems to me that much of the antitheoretical and antiempiricist rhetoric in current social theory is due to a misinterpretation of the thrust of postpositivist philosophy. Postmodern social theory calls upon postpositivism to deny the "possibility" of a scientific sociology and a general explanatory social theory. Postmodern theory feels compelled to turn into textual narrative and literary criticism because postpositivist philosophy is said to have shown that science and explanatory theory are "impossible." Sociology and science are likened to literature because postpositivism has allegedly shown that science does not have a rational method for arriving at accurate representations of reality. Hence, the postmodern argument continues, there are no important *epistemological* differences between scientific and literary knowledge. It is not that science corresponds to reality and formulates universal truths, whereas literature reflects subjective poetic imagination and depends on arbitrary aesthetic tastes. In the postmodern view, literature and science both are textual genres that use various stylistic devices and rhetorical techniques to persuade the audience of their authenticity (Habermas 1990:185ff.).

Postmodernism concludes that in a certain sense, texts are all there is.[28] In this view, the crucial difference is not that between science and literature, but between good and bad fiction. There are certain differences between scientific and literary texts, but these are differences in style and presentation, not in rationality and objectivity. Hence, the sociology of scientific knowledge turns into discourse analysis, and social theory turns into literary criticism.

I believe the postmodern equation of science and literature rests upon a serious *non sequitur*. To say that there are no epistemological differences between science and literature is not to show that there are no differences whatever. Even if we accept the critical epistemological argument that we have no reliable way of knowing when our knowledge corresponds to reality, we still do not have to accept the flawed postmodern conclusion that therefore scientific does not differ significantly from literary knowledge. For the postpositivist argument is not that science and general explanatory theory are "impossible," it is just that modern science does not need any epistemological foundations to continue doing what it has been doing since the Scientific Revolution.[29] That is, we may accept the postmodern argument that there are no epistemological differences between science and literature, but we do not have to accept the erroneous conclusion that there are therefore no differences at all and that texts are all there is. Drawing this conclusion would be like saying that since we have no way of telling whether afterlife looks more like the Christian heaven or the Buddhist nirvana, there are no interesting sociological differences between Christianity and Buddhism. The postmodern critique of positivist epistemology is deeply flawed *because it remains an essentially epistemological critique*. For in a nutshell, the postmodern argument is this: since science does not correspond to reality, it is like literature, and hence texts producing the rhetorical fiction of reality are all there is.

From an organizational perspective, however, there are considerable differences in social organization between science and literature, and among various scientific fields. Consider again figure 6.2, p. 157. The postmodern equation of sociology with literature reflects the fact that they have similar control structures, especially when compared to the densely structured research fronts in the physical sciences. Sociology and literature both have less severe entry restrictions, comparatively low levels of resource concentration, and looser collegiate controls. Lay practitioners are not entirely excluded from sociological production, and the material means for intellectual production are widely dispersed. In addition, a large number of separate reputational networks in sociology weakens collegiate control.

Current social theory, in particular, is a highly decentralized and patrimonial field. Thus, theoretical work in sociology is like literature in that it does not require expensive equipment and copresence among several practitioners. Like literature, theoretical work in sociology is more lonely mental production. Such fields tend to develop highly individualistic and discursive workstyles, and they produce conversation, commentary, and debate more than facts. The comparatively decentralized control structure allows for the coexistence of multiple approaches and perspectives, and so, the social pressures for coordinating work and adjusting one's statements to those of other theorists are fairly low. Ever since the decline of Parsonsian functionalism in the late sixties and seventies, social theory has turned into a pluralistic and very cosmopolitan community. The influx of European theories has further diversified the cultural outlook of the field. Like literature, social theory is published more in books than in articles, which reflects the low level of conceptual integration and cumulation in the field.[30] Moreover, just like literary criticism, much work in current social theory is commentary and interpretation of classical texts and their authorial intentions. Like in literature, the pictures of our most famous hermeneuticians can sometimes be found on the jackets of their books.

The organizational perspective yields a theory about why postmodernism is skeptical of scientific and general explanatory theory, and why it likens sociology to literature. Highly decentralized fields tend to produce conversation rather than facts. Facts and unified theory emerge when other scientists incorporate one's statements into their own work. Facts and scientific theory are produced by accepting statements as premises for one's own statements. In this way, statements are transformed into black boxes or building blocks for further research. To induce other scientists to turn one's statements into black boxes, scientists can mobilize a variety of textual and nontextual resources, such as Reason and Reality, laboratory equipment, networks of other scholars, reputation, and numbers. Fact production is thus more likely when these resources are highly concentrated, as is the case in research front networks.

In decentralized fields, however, these resources are more widely dispersed; and inducing other scientists to turn one's statements into black boxes is much more difficult. Allies are hard to come by. In decentralized fields, such as social theory, *texts* are the most important resource. Since collegiate control structures are fairly loose and multiple sources for gaining reputations coexist, scholars must rely to a great extent on the persuasive textual force of their arguments. This is why postmodernism stresses the role of rhetoric and literary style in sociology. In conversational fields, "good writing" becomes an important concern since nontextual resources to support claims are either nonexistent—like lab equipment in many social sciences—or widely dispersed.

There is, then, a certain plausibility to the postmodern equation of sociology with literature. But sociology, and especially social theory, are like literature in that they have similar structural characteristics as professions, not because postpositivist epistemology has shown that texts are all there is. There is not a great deal of confidence in scientific and explanatory theory in current sociology, but I think this is so because the social and organizational structures of the field are more conducive to conversation and rhetorical persuasion than to fact production and theoretical unification. Current social theory does resemble art, but not for the epistemological reason that scientific sociology is "impossible," but for the sociological reason that sociology is not a very strong collegiate profession. Postmodernism is the ideology of weak textual fields.

This argument implies that a scientific sociology is possible, but would require substantial changes in the control structures of the profession. This argument also implies that it is conceivable to have a "scientific literature" with very strong collegiate controls and high density, and to have a "literary physics" with much weaker entry restrictions and mutual dependence. It seems that Greek drama and the genre of Socialist Realism, for example, come close to "scientific literature," while seventeenth-century physics was probably much more discursive than contemporary physics. Let me finally elaborate very briefly on this latter case.

Before the academic professionalization of science in the reformed Prussian university system, physics, like other sciences, was largely the business of "gentleman-amateur" researchers (Morrell and Thackray 1977; Whitley 1984).[31] That is, physics was not yet a strong collegiate profession. The gentlemen amateurs were wealthy and educated lay researchers who conducted experiments in privately owned home laboratories. There existed few formal entry restrictions, for anyone who had the time, money, and general educational background could participate in research. Of course, in effect only members of the upper classes had the resources to do science, but these more general social entry restrictions did not create strong *professional* controls over practitioners. There was as yet no regular academic labor market for scientists. Hence, many gentlemen amateurs were autodidacts, for the preprofessional sciences lacked extended periods of professional training and credentials as necessary prerequisites for "legitimate" practitioning.

As a result, the resources for intellectual production were not yet concentrated in universities and institutionalized laboratories. The material means of scientific production were to a large part owned by wealthy individuals. Hence, instruments and work techniques were subject to a great deal of individual discretion and control. The low level of resource concentration was thus conducive to highly idiosyncratic and sometimes even incompatible ways of doing research.

Consequently, the level of collegiate control over practitioners was rather low. There existed no elaborate systems of peer review and inspection so that individual researchers enjoyed a great deal of autonomy in their work. Status in those highly informal and personalized preprofessional communities was due more to general social visibility than to shared collegiate evaluations of one's work. Since resources were widely dispersed and owned by private individuals, the preprofessional community lacked the authority to effectively exclude "illegitimate" or "unscientific" practitioners from its ranks.

As a result, preprofessional seventeenth-century physics probably looked more like contemporary social theory or even literature than like twentieth-century physics. The discursive mode of seventeenth-century physics was highly informal and controversial.

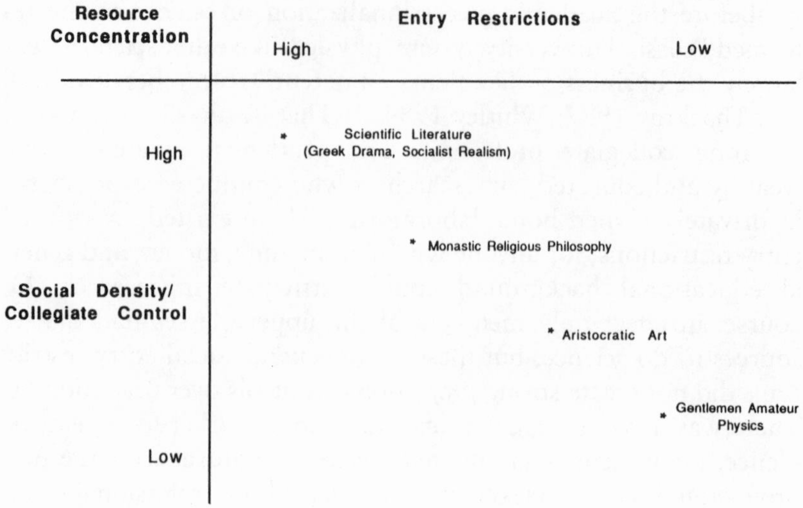

Resource Concentration	Entry Restrictions	
	High	Low

High	* Scientific Literature (Greek Drama, Socialist Realism)
	* Monastic Religious Philosophy
Social Density/ **Collegiate Control**	
	* Aristocratic Art
	* Gentlemen Amateur Physics
Low	

FIGURE 6.3   A Map of Professional Fields

Research practices were strongly influenced by idiosyncratic preferences and local customs (Whitley 1984). There was a great deal of uncertainty over the "right" ways of doing physics, and multiple approaches were able to coexist because individuals owned the means of scientific production. Since practitioners were mostly amateurs, lay audiences influenced the standards of scientific production, and so seventeenth-century physics remained embedded in the older mystical and alchemist traditions (Webster 1982). Like contemporary social theory, seventeenth-century physics was probably more of a "conversational" discipline without clear boundaries, paradigms, and shared methodologies. Decentralized control structures and dispersed resources allowed for highly individualistic workstyles and more patrimonial communication forms.

Figure 6.3 very tentatively arranges "scientific literature" and seventeenth-century physics, among other fields, in the comparative social map of professional disciplines. Note that when compared to figure 6.2, the distribution of fields is now exactly reversed: literature is now a strong, physics (i.e., "research front science" in figure 6.2) a weak profession. This reverse order clarifies my main point: differences in the "scienticity" of professional fields are not so much due to ontological differences in subject matter, or to epistemological differences in the foundations of knowledge, but to varying levels of professionalization.

SUMMARY AND CONCLUSION

In the present chapter, the technological cum neo-Durkheimian approach to science was extended to some other professional fields to demonstrate the strongly comparative and comprehensive character of the theory of scientific organizations. Task uncertainty explains variations in status and prestige in those fields where rankings depend on collegiate evaluations of one's work. The more uncertain the area of work, the more innovative the contribution to the field, and the more prestigious one's position in the profession-al status hierarchy.

On the horizontal dimension, it was found that resource concentration, entry restrictions, and mutual dependence explain differences in the workstyles of various professions. The more concentrated the means for professional production and the tighter the entry restrictions, the more colleagues will depend on each other for recognition and rewards, and the more "scientific" and "objective" the profession's cultural outlook will be.

From this perspective it becomes understandable why postmodernism and deconstructionism equate science and sociology with narrative and rhetoric. Sociology and literature have similarly weak control structures, do not establish strong collective controls over practitioners, and so are more informal and discursive. The important difference is not between professional fields *per se*, but between fields with high resource concentration, severe entry restrictions, and tight mutual dependence versus fields scoring low on these variables. This is why seventeenth-century physics differs from modern physics.

CHAPTER 7

A Theory of Scientific Production

In this chapter, I shall pull together the various systematic arguments made throughout this book about the relationships between social structural arrangements and the discursive modes or cognitive styles of social groups and professional communities. In its present version, the technological cum neo-Durkheimian theory of science is rather incomplete and schematic. First, I have not accounted yet for variations *within* various natural and social sciences. The popular ontological and epistemological distinctions between the social and natural sciences not only miss those significant internal differences between scientific fields that cut across the social versus natural dichotomy, they also tend to disregard historical changes in the social and intellectual organization of these fields. From an organizational perspective we would expect, for example, that some social sciences look more like our familiar image of the "hard" and "mature" natural sciences than other social sciences. We would also expect seventeenth-century physics to differ quite extensively from contemporary physics. In the area of general organizations, Charles Perrow (1972) makes a similar point when criticizing conventional schemas for distinguishing between private and public, profit and nonprofit, or between industrial and nonindustrial organizations. The crucial distinctions, says Perrow, do not revolve around types of organizational goals, services, clients, or relevant environmental sectors, but around the nature of the work done in organizations. Some churches look more like factories, and some schools look more like welfare agencies. Correspondingly, scientific organizations do not differ because they research different areas of reality, as is assumed in the traditional social versus natural dichotomy, but because they do their work in different ways.

Second, the preliminary model developed in chapters 4 through 6 needs elaboration because its two main variables, task uncertainty and mutual dependence, cannot plausibly be left unexplained as exogenous variables. In fact, both variables require systematic and historical explanation. Systematically, there are excellent general sociological reasons to assume that mutual dependence, for example, varies with such factors as group size and the concentration of resources for scientific production. And historically, neither task uncertainty nor mutual dependence can be expected to remain constant over time. Task uncertainty, for example, is not simply given with the ontological matrix of a particular scientific domain or subject matter, but rather follows from the organizational and instrumental ways in which such domains are collectively researched. For the most part, tasks are not certain or uncertain in and of themselves; what is certain or uncertain are the ways in which tasks are socially perceived and work is collectively organized. This is the point of W. R. Scott's (1981:231) lucid critique of the one-sided determinism in some technological theories of organizational structure.

The formulation of the theory of scientific organizations in this chapter draws rather extensively upon two major organizational theories of science: Randall Collins's *Conflict Sociology* (1975:470–523), and Richard Whitley's *The Intellectual and Social Organization of the Sciences* (1984). Whitley's argument is an extension of Collins's but, unfortunately, Whitley is occupied more with comparative statics and conceptual classification than with explanation and causal dynamics. In his core chapter 5, Whitley designs a typology of sixteen "logically possible" scientific fields, of which only seven are said to be "empirically stable." Whitley then goes on to show how actual scientific fields fit into these structural types, ranging from "fragmented adhocracies" over "polycentric professions" to "conceptually integrated bureaucracies."

Such exercises in comparative statics are useful for descriptive and analytic purposes, but they tend to be insensitive to the dynamics of scientific change. The historical evidence on scientific fields and specialties shows that disciplinary boundaries are constantly changing, that certain research specialties overlap considerably with other specialties, that frequent scientific migration creates new fields, and that older disciplines change their structural outlooks in response to technological and organizational innovation. The notorious problems sociologists of science report when trying to identify the precise boundaries of scientific fields, to determine who is a member of a specialty network and who is not, or to map the continuous branching of established fields into new areas of ignorance provide ample evidence for the problems with classifying sciences according to structural types. Daryl Chubin (1983:8) lists no less than fifteen commonly used operationalizations of scientific fields, such as "networks," "research areas," "invisible colleges," and "paradigmatic groups." This operational diversity reflects more than just a methodological problem: depending on how scientific fields are operationalized, they belong to this rather than to that organizational "type."

As a result, assigning scientific fields to structural types yields an overtly static picture of science. At best, the characterization of a field as a "polycentric profession" or a "fragmented adhocracy" is likely to be outdated rather quickly.[1] The sciences are self-referential systems of knowledge production (Luhmann 1984), meaning that no one but scientific communities themselves can negotiate and decide what topic areas to include in a given field, who is a member of a specialty network, what counts as a significant contribution to a cluster of problems, and how and when disciplinary boundaries will be drawn and redrawn. Even citation and cocitation measures of scientific communication do by no means unambiguously identify disciplinary boundaries and specialty networks, although these measures do assume the internal perspective of participating scientists (see Gilbert and Woolgar 1974). Typologizing scientific fields from an observer's point of view, then, is a somewhat futile effort.

Also, the logical structure of Whitley's argument remains opaque. He spends too much time distinguishing between types of variables, such as "functional" and "strategic" dependence (p.88), or "technical" and "strategic" uncertainty. But due to my interest in cumulative theory, I have adopted some of Whitley's variables, added some others, and tried to present the argument in a more straightforward form. But most importantly, I believe that Whitley's argument falls short of realizing the full potential of an organizational theory of science: explaining the *contents* of knowledge, in the sense of discursive styles.

The theory of scientific organizations is represented in figure

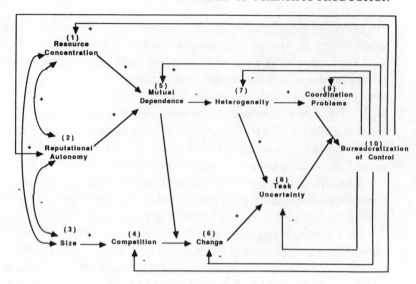

FIGURE 7.1   A Theory of Scientific Production

7.1. In my discussion I shall focus on what I hold to be the most important causal relationships. Also, I shall concentrate on the feed-forward effects, since these stand for my basic argument that variations in discursive practices are to a large extent determined by the structural arrangements in scientific organizations. However, the feedback arrows indicate the nondeterministic character of the theory.

RESOURCE CONCENTRATION

Resource concentration indicates the patterns in which the material and symbolic means of scientific production are distributed throughout communities and specialty networks. Material means include status positions, research facilities, publication outlets, and grants, while the most important symbolic resources for scientific production are peer recognition of one's work, as in citations and awards, and the reputations that accumulate recognition. Resources are highly concentrated when small but powerful groups of researchers control access to the material and symbolic means of scientific production.

Resources tend to be highly concentrated in fields using very expensive equipment. Such fields are also highly stratified

(Traweek 1988). There exists a clear prestige hierarchy in journals, awards, academic departments, and laboratories (Cole and Cole 1973), and the gatekeepers who control the most prestigious resources constitute the elite cores of such fields (Zuckerman 1977). Also, such fields exhibit a very clear prestige ranking of various subfields and specialties. Some fields are considered more significant and important than others. The most prestigious scientists work at the research fronts in the innovative and cumulative core areas of a discipline, and they define where the field as a whole is moving. Whitley (1984:201ff.) calls such fields "conceptually integrated bureaucracies," for they have rather visible prestige hierarchies among practitioners, facilities, publications, departments, and specialties. Modern physics appears to represent this pattern rather closely.

Conversely, resource concentration tends to be low in fields where research facilities and resources are less expensive and more widely distributed across communities and specialty networks. In such fields the control and communication structures are more decentralized, for there exists a wide variety of material and symbolic resources for intellectual production. Fields with low resource concentration provide for diverse and widely distributed sources for scientific production so that a large variety of publication outlets, research facilities, and reputational networks are controlled by groups that are, to an extent, independent from each other. Accordingly, fields with decentralized resource control do not exhibit clear and agreed-upon patterns of internal stratification so that there exist no obvious prestige rankings among subspecialties, journals, and types of intellectual activity. There will be rather intense conflicts between various groups of practitioners over the relative importance of their particular specialty for the field as a whole. However, if resource concentration is extremely low, then there will be very few evident prestige rankings among resources.

In sociology, for example, it would be hard to tell whether gerontology is "more significant" for the field as a whole than, say, family sociology, and it would probably be argued that such questions are meaningless to begin with. That is, fields with low resource concentration and stratification have more egalitarian and pluralistic control and communication structures, and they lack the cumulative and rapidly advancing research fronts found in fields with highly concentrated resources.[2] In current sociology, for example, there exist many regional associations, specialty journals, separate collegiate networks, and reputational elites that sustain a variety of coexisting approaches and perspectives so that the field is more fragmented than cumulative (see Turner and Turner 1990).

REPUTATIONAL AUTONOMY

According to Whitley (1984:220ff.), reputational autonomy represents the extent to which professional groupings can proceed self-referentially in allocating recognition, rewards, and reputations for scientific contributions. That is, reputational autonomy measures the extent to which particular scientific organizations are the sole source for academic recognition and reputation. Reputational autonomy corresponds to the level of professionalization in scientific organizations. High reputational autonomy exists in fields with rather severe formal and informal entry restrictions. If a scientific organization and its dominant groupings control the educational and credential systems, and if professional training and credentials are prerequisites for obtaining employment and access to research facilities, then formal entry restrictions are severe, and the reputational organization is highly professionalized. Formal entry restrictions exclude nonprofessional practitioners from contributing to "serious" science *de facto*, though not *de jure*. Severe formal entry restrictions, or advanced professionalization, increase reputational autonomy because only "established" scientific organizations may determine who is a "legitimate" scientist formally qualified to make "serious" contributions to knowledge (see Collins and Pinch 1982).

In highly professionalized scientific organizations with high reputational autonomy, formal entry restrictions are reinforced by more informal entry restrictions. Informal entry restrictions reflect the disciplinary structures of particular fields and the interrelations between them. Informal entry restrictions are severe when the boundaries separating fields and specialties from one another are rather impermeable. For example, fields using highly esoteric symbolisms and techniques, very specialized research equipment, and rather exclusive publication outlets make it virtually impossible for "outsiders" to contribute to or even understand knowledge. Within sociology, the subfield of mathematical sociology clearly follows this pattern. As a result of impermeable boundaries, the groupings controlling the reputational networks within such fields are very powerful and autonomous in allocating recognition and resources. In sum, reputational autonomy is high in strongly professionalized scientific organizations that couple academic contributions and recognition to severe formal and informal entry restrictions.

Conversely, reputational autonomy is low when scientists may obtain recognition and rewards from a large variety of organizational sources. In this case, any particular reputational network is not the only possible source for recognition, and scientists may obtain credits for their contributions from multiple sources. Typically, such fields have rather transparent disciplinary boundaries, and there are many schools and independent specialties. Weak disciplinary boundaries permit practitioners from different but related fields to make contributions to the focal field, as is the case in current social theory where other social scientists, literary critics, and philosophers figure quite prominently. Often, loose disciplinary boundaries are even celebrated for opening up "broader" perspectives and "multidimensional" approaches, and there are persistent calls for "interdisciplinarity" (Giddens 1984; Calhoun 1988).

Most importantly, fields with low levels of disciplinary professionalization, weak formal and informal entry restrictions, and loose organizational boundaries are often influenced by lay audiences, as is the case in many areas of "applied" research in sociology, but also in biomedical research and much of industrial science. In this case, lay audiences are an important resource for recognition, and they may also influence the general standards and directions of research. In certain areas of "interpretive" or "qualitative" social research, for example, lay members are even called upon to validate professional research findings ("member validation").[3] Lay audiences provide an alternative nonacademic source of recognition and rewards, and sometimes recognition of one's work by the general public is even considered a particularly prestigious form of reward. The "Footnotes" published by the *American Sociological Association*, for example, regularly announce sociologists' works being mentioned in the mass media.

Reputational autonomy, then, indicates how the reward systems of scientific organizations are controlled. The more particular professional groupings monopolize the distribution of reputations, and the more difficult it is for "outsiders" from other fields or for lay audiences to make competent contributions to a field, the more autonomous are the established practitioners in defining the standards for proper research and in awarding reputations for serious and competent contributions. Conversely, practitioners in fields with low reputational autonomy, weak disciplinary boundaries, and close relations to lay audiences are able to obtain recognition from a variety of professional and nonprofessional sources, and, as a consequence, no particular group exclusively controls the reputational system. Reputational autonomy can be expected to covary positively with resource concentration, for those core groups who control the means of scientific production are also likely to control the reputational system, and vice versa.

MUTUAL DEPENDENCE, HETEROGENEITY AND COORDINATION PROBLEMS

As was already discussed in the previous chapter, mutual dependence indicates the degree to which scientists are dependent on their peers in gaining recognition for their work, reputations, jobs, and access to research facilities. Mutual dependence is high when practitioners depend very closely on particular peer groups, and it is low when multiple peer groups grant equally valuable reputations in a largely autonomous way. We hypothesize that mutual dependence (5) is determined by resource concentration (1) and reputational autonomy (2). When the means of scientific production are highly concentrated, and when there are not a lot of alternative sources for gaining recognition and reputations, then scientists depend very closely on the particular peer groups who control these resources. Conversely, when resources are widely dispersed throughout a community, and when weak formal entry restrictions and transparent disciplinary boundaries enable scientists to gain reputations from a large variety of peer and, possibly, lay audiences, then researchers have a wider range of options for doing and obtaining rewards for their work.

However, even if there exists a variety of sources for recognition and reputation, mutual dependence might still be rather intense. This is so because not all these sources are equally prestigious. Fields with high resource concentration and reputational autonomy are internally stratified in that various subfields and specialty areas are accorded varying amounts of prestige and overall disciplinary significance. That is, some fields are considered more "important" and "significant" for the field as a whole than others, and the most prestigious subfields are usually the highly uncertain, innovative, and cumulative core areas of the discipline. In this case, practitioners do have options to work in a number of less visible subfields, but will tend to compete for entry into the most prestigious core areas. That is, mutual dependence may remain high even if a variety of subfields offer alternative work opportunities, for practitioners will still depend closely on the peer groups who control those core areas.

Mutual dependence is expected to determine the level of cognitive integration within a field. When scientists depend very closely on each other for recognition and reputation, they are more likely to adjust their work practices and cognitive frames to each other. Mutual dependence, or social density, increases the "group" aspect of scientific communities and networks. That is, a high level of mutual dependence will tend to generate a strong sense of collective identity, a prominent emphasis on group conformity, and a confident belief in the group ways of doing research. Group pressures induce scientists to coordinate their work with one another and, as a result, will create rather uniform and shared research practices. The group watches closely over the proper standards for doing good science, and there is not much tolerance for individual deviance and innovation. Of course, group controls and pressures are tempered by a fair amount of individual discretion in dealing with complex and uncertain problems, especially in the innovative core areas of a field. But generally, high levels of mutual dependence or collegiate control tend to generate rather uniform and shared symbolisms, research practices, and paradigmatic approaches. As a result, the coordination of research styles and results is facilitated, for shared practices and common scientific symbolisms permit smooth communication among scientists.

Conversely, under conditions of low mutual dependence the

cultural outlook of a field will be rather heterogeneous. For low mutual dependence implies that scientists and scientific networks are rather autonomous in designing research strategies and communicating results. Under conditions of low mutual dependence, the group pressures on adjusting one's work to that of other scientists are much weaker, and hence scientists and scientific groups are largely independent from one another. As a result, the field will consist of a variety of separate schools, approaches, and perspectives that lack overall conceptual integration. While the level of mutual dependence might remain high *within* these schools and clusters, the overall unity of the field is very weak. Hence, coordination problems will be severe, for the separate research schools and intellectual traditions will find it difficult to communicate in terms of common and shared scientific symbolisms. Autonomous schools and research networks are free to develop distinct scientific vocabularies and research techniques.

SIZE, COMPETITION, AND CHANGE

The important effects of size on social organization have long been recognized by the sociological tradition. Spencer (1898), Durkheim (1893/1965), and Simmel (1890) agreed that demographic growth leads to competition over scarce resources, which in turn leads to increasing social specialization and differentiation. In contemporary social research, size still figures prominently in the study of formal organizations, and it has repeatedly been observed that size increases structural differentiation within organizations (e.g., Pugh, Hickson, Hinings, and Turner 1968, 1969). Size is defined here as the number of formally qualified scientific practitioners relative to the available amounts of jobs and research facilities in a given field. All other things being equal, when size increases resources become scarce, and scientists will compete intensely over jobs and facilities, especially in the most prestigious areas of a discipline. The sociological tradition assumed that increasing competition leads to social differentiation and occupational specialization, but we need to distinguish carefully between various forms of scientific change that increasing competition will trigger under various circumstances (see Collins and Restivo 1983b).

The assumption that the effects of size and competition on scientific change are contingent upon varying structural arrangements in scientific fields is represented in figure 7.1 by the *interaction effect* of competition (4) and mutual dependence (5) on scientific change (6). Depending on the level of mutual dependence between scientists, competition will trigger various forms of scientific change: cumulation, specialization, fragmentation, and migration.

Cumulation

Under conditions of very high mutual dependence between scientists, competition among practitioners will lead to cumulation; i.e., to rapid advances in knowledge. In this case, scientists compete by making innovative contributions to highly uncertain fields of knowledge. This pattern of change by innovation and cumulation occurs in what we have called, following Price (1986), the very prestigious research fronts in a field. These small elitist networks institutionalize high levels of mutual dependence, for practitioners are closely integrated and controlled by their peers. Change assumes the form of cumulation because competition over innovations is strictly tempered by high mutual dependence so that innovations must be recognized and validated by small core groups of scientists.

That is, the high level of mutual dependence prohibits innovations from diversifying and disorganizing the cognitive practices and work routines of research front communities. Change does occur because competition induces innovations, but the close integration of small core networks assures that change occurs as orderly cumulation. We would expect priority disputes and discovery conflicts to be most intense and frequent in these core areas, for competition *is* competition over innovations.

Specialization

If mutual dependence is intermediate, however, or if size increases are so dramatic that the majority of scientists will not be able to work in the innovative and cumulative core areas in a field, then change will be likely to occur as specialization. "Specialization" is understood here according to Kuhn's (1970) description of normal science. Most importantly, specialization occurs as the application of a particular paradigmatic framework to new problem areas. Scientific contributions do not so much consist of innovative modifications of a paradigm, or of dramatic "discoveries," but rather of more routine expansions of established paradigms to special problems within a field. Hence, specialization creates a variety of subfields that are distinguishable in terms of their problem domains and subject matters and that are also ranked in terms of their "significance" for the field as a whole. However, the fairly high amount of mutual dependence prohibits specialization from leading to the disintegration of research practices.

That is, the various subfields and specialties remain to a large extent embedded in larger disciplinary frameworks, and they are not autonomous enough to become separate reputational organizations with distinct or even incompatible research practices and paradigmatic approaches. Specialized fields thus remain integrated, for specialties differ more in terms of their specific problem areas than in their overall approaches to their subject matters.[4]

Fragmentation

Under conditions of low mutual dependence, however, change will occur as the fragmentation of a discipline. In this case, competing scientists and groups will be able to establish separate reputational networks with largely autonomous control over the symbolic and material means of scientific production. In such fields there exists a variety of schools and paradigms that are largely independent from one another and that create their own organizational infrastructures. The overall disciplinary organization is not strong enough to maintain a prominent sense of collective identity, and there are too many subspecialties, approaches, and methodologies to discern some shared paradigmatic practices.

For example, the proliferation of regional and specialty associations and journals in post-World War II U.S. sociology has largely fragmented the discipline, in that there are no unified and universally accepted problem definitions, methodologies, and theoretical approaches. Practitioners in various specialties are largely independent from practitioners in other specialties so that there is organizational room to develop very distinctive, and sometimes incompatible, ways of doing sociology. It seems that the fragmentation of sociology occured in response to increases in size and competition (Turner and Turner 1990).

Migration

The migration of scientists into new areas of ignorance is always an option in increased competition, but appears more likely in fields with comparatively high levels of mutual dependence. For if mutual dependence is very low, such as in many social sciences and humanities, scientists are free to establish their own subspecialties, such as "feminist sociology," or the "sociology of emotions," while remaining in the larger discipline. In other words, if mutual dependence is low, fragmentation seems to be the most advantageous and least risky strategy to deal with increasing competition, and to establish autonomous reputational networks with vested interests in particular subspecialties.

This is so because migration is generally a high-risk strategy. Migration amounts to building new reputational organizations, and this requires symbolic and material means for scientific production. Also, the legitimacy of such a new organization will be strongly contested by established reputational networks. Establishing a new reputational organization requires a great deal of resources, such as departments, status positions, publication outlets, specific disciplinary symbolisms, distinct approaches, and the like. New reputational organizations will have to compete with established fields over funding, students, and the definitions of problems.

That is, migration is likely to occur when cumulation—or access to the innovative core areas of a field—and specialization cannot be pursued, and when the level of mutual dependence is too high to allow for fragmentation of the discipline. Migration is especially likely when scientists have worked in low-prestige fields. And the more specialized a field becomes, the more likely it is that specialties do not carry a lot of overall disciplinary prestige and significance. That is, migration will occur when the possibilities for further specialization of an otherwise integrated field are exhausted, or when these newly created specialties decline in prestige.

TASK UNCERTAINTY

As was discussed in the previous chapter, task uncertainty indicates the extent to which scientific problems are perceived as complex and unpredictable. Task uncertainty is high when there are no standard and agreed-upon ways of dealing with a problem, and it is low when more routine problems can be solved by more predictable and generally accepted research techniques. Generally, task uncertainty is high whenever complex problems require innovative approaches, and when there is not a great deal of agreement over the appropriate ways of solving a problem. Conversely, task uncertainty is low whenever routine problems may be solved by standard approaches, and when there is not much controversy and disagreement over the right ways of doing research.

The basic argument here is that task uncertainty does *not* primarily follow from the intrinsic complexity of particular subject matters, but from the patterns of social organization of scientific fields. In figure 7.1, it is hypothesized that task uncertainty (8) follows from heterogeneity (7) and change (6). The more heterogeneous a particular field, the more disagreement exists between the various schools and perspectives, and the less it can be claimed that any particular approach is superior to all the others. Task uncertainty is increased by competing schools and approaches debating the right ways of doing research. Conversely, a high degree of agreement and uniformity in more closely integrated fields will generally reduce task uncertainty, for there are generally accepted standards for doing research.

Also, task uncertainty is higher in fields with high rates of change, no matter whether change occurs as cumulation, specialization, fragmentation, or migration. However, it can be argued that task uncertainty is lowest in fields with specialization as the dominant mode of change, for specialization requires only the expansion of established research practices into new problem areas, not the innovative modification or disintegration of paradigms themselves. In fields with high rates of innovation and cumulation, and in fields with fragmented disciplinary approaches, task uncertainty will be higher than in fields which subspecialize on particular tasks while maintaining their overall paradigmatic integrity.

BUREAUCRATIZATION OF CONTROL

"Bureaucratization of control" should be understood here not in the narrow sense of the Weberian ideal type, but rather as the level of standardization and formalization of research practices and modes of communication. Scientific production is always uncertain, and this comparatively high level of task uncertainty limits the extent to which control over scientific production can be bureaucratized. Bureaucratization of control, then, refers to the ways in which research is organized and findings are communicated. Control is highly bureaucratized when research practices follow more standard methodological canons, when findings can be expressed and assessed by rather formalistic criteria of evaluation, and when colleagues can measure the merits of contributions rather routinely. These are the "fact producing" fields of chapter 4. On the other hand, the level of bureaucratization of control is low when research practices follow more idiosyncratic and localistic standards, when there is a great deal of variation in research findings, and when colleagues disagree sharply over the relative merits of scientific contributions. These are the conversational fields discussed in chapter 4.

In figure 7.1 it is assumed that the level of bureaucratization of control (10) is due to an *interaction effect* between task uncertainty (8) and coordination problems (9). This interaction effect expresses the assumption that research efforts will be coordinated bureaucratically *only* if task uncertainty is low (Thompson 1967). That is, standard methodologies, routine assessments of results, and regular research practices will coordinate activities only if the tasks involved are not too uncertain. These fields have high group consciousness, a strong confidence in the value of "scientific" methodology, and a prominent emphasis on rigorous standards for doing and evaluating research. Such fields are perceived as very "mature," for they produce "scientific" and "objective" knowledge (Fuchs and Turner 1986).

Conversely, under conditions of high task uncertainty, coordination problems will be solved more by informal adjustments and ad hoc mutual consultations. This prediction corresponds to the arguments and findings of the technological school in organization studies introduced in chapter 5. High task uncertainty implies that there are no rigid standards for doing proper research, that what counts as "good science" is ambiguous and controversial, and that there are only few routine solutions to problems. High task uncertainty is found in the innovative core areas in a field, where science is science-in-the-making rather than ready-made science (Latour 1987). High task uncertainty is also found in fragmented fields with low overall disciplinary integration, for various approaches and schools compete here over the right ways of doing research. Under these conditions, scientific research activities will not be coordinated bureaucratically but informally; that is, according to contextual and situational negotiations.

SUMMARY AND CONCLUSION

In the present chapter, the various neo-Durkheimian, materialistic, and organizational arguments made throughout this book were integrated into a general theory of scientific production. Again, differences in the degrees of scienticity and objectivity between various fields result from differences in structural organization, not from ontological differences between subject matters. Very mature and scientific fields are those controlled by highly professionalized and self-referential organizations that have high levels of social density or collegiate control. Conversely, soft and immature fields are those which lack fully professionalized control structures, and which are internally disintegrated.

The theory of scientific organization addresses the contents of science without adopting the excessive situationalism and casuism of microsocial studies of science. This theory also incorporates the Mertonian research on scientific community organization and on behavior patterns of scientists, such as migration or fragmentation. The theory of scientific organizations also relates to the general sociology of knowledge because it treats scientific communities like ordinary, albeit very powerful, social groups. The authority invested in scientific knowledge is ultimately due to the strength of the organizations that produce it.

CHAPTER 8

Hermeneutics as Deprofessionalization

In this final chapter, the analysis moves from more theoretical concerns to the detailed investigation of a concrete empirical case; that is, the present state of sociology. There has been much debate over the nature of the field, its relationship with the natural sciences, its basic assumptions and concepts, its relative preparadigmatic immaturity, and the like. Usually, these debates are conducted as foundational controversies and, as such, have proven to be irresolvable. I shall approach these issues from an organizational angle, and so try to fulfill the reflexivity postulate of the Strong Program: a sociology of science must include a sociology of sociology.

Ever since their emergence as academic fields, especially since the classical *Methodenstreit*, the social sciences have been involved in a debate over their ontological and methodological status. This debate has been fought under such various headings as idiographic versus nomothetic methods, understanding versus explanation, hermeneutics versus science, qualitative versus quantitative research, lifeworld versus system, or agency versus structure (see Dallmayr and McCarthy 1977). The dualistic position in this debate suggests that the universe consists of two different types of entities, such as reasons and causes, mind and body, spirit and matter, action and structure, or people and things. The dualistic or hermeneutic argument is that these different entities require distinct methods for studying them so that social scientific methods differ from those employed by the natural sciences. Conversely, the unitarian or positivist view insists on the unity of the world and, consequently, suggests only one unified logic of scientific research. According to this view, the social sciences are best advised to adopt the methods and epistemology of the more mature natural sciences. Positivism holds that there is an independent reality of social facts "out there" that can be discovered and explained by scientific methods and laws. The symbolic nature of social worlds does not stand in the way of naturalizing society and subjecting its dynamics to some form of "social physics."

In sociology, much of the current methodological discussion strongly resembles the older arguments initiated by Dilthey (1883) and Weber (1949). That is, old and new hermeneutics propose a set of ontological and methodological dualisms between social and natural worlds, and between the ways in which social and natural sciences study their respective worlds (Giddens 1976, 1984; Habermas 1984; Taylor 1971; Apel 1981). Ontologically, social and natural worlds are assumed to differ in that the social world, but not nature, is constituted symbolically. Social reality is structured symbolically in that members interpret their own actions and gestures and communicate their interpretations to others. Nature, on the other hand, is sheer physical and observational reality because nature does not talk. Therefore, the dualistic hermeneutic argument continues, the methodology appropriate for the study of nature is inadequate for the study of society. Whereas natural science "interprets" only insofar as the internal relationships between paradigms, theories, and methods require conceptual explication, social science always involves a "double hermeneutic," for the objects of social science are subjects who interpret themselves before the researcher does:

> The social scientist studies a world, the social world, which is constituted as meaningful by those who produce and reproduce it in their activities—human subjects. (Giddens 1982:7)

In sociology, then, hermeneutics is generally understood as a unique *method* for interpreting symbolically constituted social worlds. As interpretive or qualitative method, hermeneutics is contrasted with quantitative or positivist method. Interpretivists intend to study social worlds from the inside perspective of members and generally prefer "soft" and inductive methods and case-study approaches in natural settings. Positivists, on the other hand, study social worlds from the outside perspective of science and generally prefer "hard" quantitative or comparative methods that can back up larger-scale generalizations. Interpretivists are more skeptical of science's superior cognitive status and point out that ordinary members of social groups are competent cointerpreters of the social world. Positivists generally think that the rational and critical structure of scientific method places the scientist in a superior cognitive position not available to lay people. Thus, like its romantic predecessor, the current sociological *Methodenstreit* is mainly a debate on methodology, nourished by ontological presuppositions about the fundamental differences between social and natural worlds.

In philosophy, on the other hand, the discussion on hermeneutics has moved away from the older ontological and methodological issues (Fuchs and Wingens 1986). In philosophy, hermeneutic understanding is understood primarily not as method but as the practice of Being itself (Heidegger 1962), as the normative self-understanding of cultural tradition (Gadamer 1975), and as the medium of discursive edification (Rorty 1979). Although several diverse approaches exist in philosophical hermeneutics, the common Gadamerian point seems to be that hermeneutics is what we obtain if we abandon the very idea of methodology.

Adopting this antimethodological stance of philosophical hermeneutics, I want to suggest that *Verstehen* is not the specific method used by the social sciences for ontological reasons. Rather, I believe that hermeneutics refers to the "life-form"[1] we establish when we decentralize and democratize the control structures of sociology as a profession. Whereas the traditional debate on the ontological and methodological status of the social sciences has concerned mostly social philosophers and philosophers of science, I intend to turn this debate into a *sociological* argument by suggesting that the current *Methodenstreit* has its underpinnings in organizational politics and structure.

This implies that I do not have to offer any *philosophical* solutions to the current *Methodenstreit* in sociology. I do not propose any answers as to whether social worlds "really differ" from nature, and whether these "real differences" necessitate different methodologies for studying social and natural worlds. Likewise, I do not claim to have a metaphysical criterion that would tell us whether hermeneutics or positivism is closer to the truth. I shall not deal with the question of whether interpretivism or positivism is the superior way of studying social reality, and so I do not evaluate the rationales interpretivists and positivists rely upon in their respective approaches to the social world. Correspondingly, my critique of interpretivism does not in any way intend to question the value of, say, ethnographic fieldwork or unstructured interviews. Neither do I suggest that positivist studies with a lot of numbers, coefficients, and significance tests are in some way better or more realistic than a participant observation study or an oral history. With respect to the truth claims of interpretive and positivist perspectives, my analysis follows the recommendation of the Strong Program in the sociology of scientific knowledge (Bloor 1976:4) to be "cognitively indifferent."

What I do want to point out, though, is that the metaphysical and methodological debates in sociology have sociological or, more precisely, organizational underpinnings. We might not be able to resolve the metaphysical problems of the current *Methodenstreit*, and it seems to me that the postmodern turn in philosophy (Magnus 1988) expresses the skeptical sentiment that metaphysical questions of this kind cannot possibly be answered. But as sociologists we can point out that philosophical problems have a social-structural dimension to them. While postmodern philosophy has taken the apparent irresolvability of its grand but homemade metaphysical problems as reason to turn into literary criticism, deconstruct its own narratives, and proclaim the "end of philosophy" (Rorty 1989), it seems to me that a more fruitful response is to turn grand philosophical into modest sociological problems. In this sense, the following discussion continues the Wittgensteinian attempt to develop a "sociological philosophy" (Bloor 1983).

I shall use the term "hermeneutics" in two different but related senses; one more specific and one more general. In a more specific sense, hermeneutics refers to the interpretive way of doing sociology advocated by certain groups within the discipline. In this more narrow sense, hermeneutics covers the interpretive paradigm as a specialty in our field. This specialty grants lay people some control over sociological knowledge, and so I shall argue that the interpretive paradigm corresponds to an organization with low reputational autonomy or weak external control. In a more general sense, however, hermeneutics characterizes the self-understanding of sociology as a whole. Hermeneutics then expresses the discursive, informal, and controversial mode of intellectual exchange in the overall field. In the latter part of this chapter, I shall argue that this mode is due to high task uncertainty and structural fragmentation, or weak internal control. In both specific and general senses, then, hermeneutics is what we obtain under certain organizational conditions, regardless of ontology and methodology.

THE INTERPRETIVE PARADIGM IN SOCIOLOGY

Interpretivism is by no means a unified and monolithic movement, and considerable disagreement exists between various groups of practitioners. There are differences regarding the proper techniques of research, the extent to which sociologists should get involved in natural lifeworlds, and the compatibility of qualitative and quantitative strategies. Reviewing these differences, Bryman (1984:77–78) nevertheless concludes that for interpretivism, "the *sine qua non* is a commitment to seeing the social world from the point of view of the actor, a theme which is rarely omitted from methodological writings within this tradition." Interpretive approaches such as symbolic interactionism, fieldwork ethnography, discourse analysis, participant observation, or oral history claim that social reality is socially and symbolically constructed. Members are competent, knowledgeable, and capable agents who negotiate definitions of situations, engage in social practices to account for their actions, and communicate their interpretations to others. The interpretive paradigm holds that sociology has no privileged access to its subject matter but confronts competent first-order lay interpretations of the social world. Possibly, lay interpretations might even be in a superior epistemic position, for sociologists are usually not directly involved in the social worlds and practices they study. Professional sociologists enter their domains "from the outside" (Blumer 1969), whereas lay interpretations have the cognitive advantages of "inside" accounts. Therefore, the interpretive paradigm claims—unlike natural scientists who do not encounter competent interpretations of the world they study—social scientists must somehow relate their interpretations to lay interpretations.[2]

Sociological paradigms differ in the ways they deal with these lay interpretations. Roughly, three broad options are available. The classical behaviorist paradigm suggests that we should ignore lay interpretations, for symbolic meanings belong to the black box of human cognition and, hence, cannot be observed scientifically. Theodore Abel's (1953) classical and influential critique of understanding and hermeneutics, for example, argues that since interpretation deals with such nonobservables as intentions and reasons, hermeneutics can only be useful as a heuristic device. Interpretation might generate certain interesting hypotheses about human behavior, but it cannot itself validate these hypotheses. Validation remains the task of explanatory science testing causal hypotheses.

As opposed to behaviorism, mainstream sociology, following Durkheimian (1895/1982) positivist epistemology, believes that internal and symbolic entities such as reasons and cognitions are accessible to the researcher, and that they are important "data sources." Like any other social fact, motives and intentions are part of objective reality, can be addressed by normal research techniques and are matters to be explained or explanatory variables themselves, just like any other variable. In this view, lay interpretations are responses to professional questions and do not pose any special methodological problems. The fact that lay members interpret their own actions and situations does not give reason to assume that social worlds differ fundamentally from nature. The Durkheimian survey researcher passes through lay responses as a *detour* to get to the underlying reality of social facts. It is this underlying reality the Durkheimian is really interested in, not lay interpretations *per se*. These are important only as "indicators" and "observed scores" that point at the deeper reality of facts and true scores.

The interpretive paradigm decides to take lay interpretations very seriously, although this is done in many different ways.[3] Lay interpretations are not just another set of social facts. Rather, facts exist only insofar as they are socially and symbolically constructed in processes of lay interpretation and negotiation (Douglas 1970a:11). Lay interpretations are *constitutive* of social worlds because social reality exists only *as* symbolically constructed reality.

The interpretive paradigm points out that positivism, in its attempt to model the social after the natural sciences, fails to see that unlike nature, social reality exists only insofar as lay members *create* that reality in meaningful interaction. Lay interpretations are not just measures that indicate some other underlying reality, they *are* the reality. There is no such thing as a deep reality of social facts that somehow lies underneath the reality of lay interpretations. In a sense, those interpretations are all there is, for reality is constructed in processes of meaningful interaction and accounting. And these lay interpretations might employ very different frames of meaning and reference than the professional interpretations used by the sociologist to reconstruct social reality. The meanings of lay interpretations are not necessarily the same as the meanings of professional discourse. In short, "qualitative methodology advocates an approach to examining the empirical social world which requires the researcher to interpret the real world from the perspective of the subjects of his interpretation (Filstead 1970a:7)."

That is, to a certain extent, sociological interpretations must be *consistent* with lay interpretations, for only interpretive consistency prevents sociology from simply imposing its accounts on members' accounts (Schutz 1967). Inconsistency between lay and sociological accounts results in systematically misunderstanding the distinctively symbolic and constructed nature of social reality. Therefore, the interpretive paradigm continues, we cannot simply ignore members' interpretations or treat them as facts or indicators of facts. Instead, we must accept members' definitions of reality as the *premises* or grounds on which to build our own professional definitions (Glaser and Strauss 1967). To regard lay interpretations positivistically as indicators is to start with professional constructs, and then proceed deductively toward objective social reality through the detour of lay responses. To regard lay interpretations hermeneutically as the *premises* of sociological accounts means to start with those interpretations, and then move inductively toward sociological concepts *and back*.[4]

The interpretive consistency between lay and sociological accounts is held to require *Verstehen* or interpretive understanding. That is, members' interpretations must not be "subsumed" under the categories of professional discourse, but must be understood "from within" (Geertz 1973) and by means of "sensitizing concepts" (Blumer 1954). "Sensitizing" are concepts that are close to members' perceptions of reality, and which are constantly revised in order to maintain this closeness.

This hermeneutic closeness is established through interaction.

The interpretive paradigm asserts that *talking about* social worlds requires, to some extent, *communicating with* their members. Understanding social worlds from within requires some *participation* of the researcher in ongoing processes of interaction between members (Emerson 1983; Maranhao 1986; Skjervheim 1974). In a certain sense, researchers must become members of the lifeworlds they study. Even if qualitative researchers do not chose participant observation as their method, the underlying philosophy of qualitative method points at the necessity to understand members' lifeworlds from within:

> In qualitative methods, the researcher is necessarily involved in the lives of the subjects. . . The researcher must identify and empathize with his or her subjects in order to understand them from their own frames of reference. (Bogdan and Taylor 1975:8)

That is, qualitative methods emphasize members' points of view and argue that a detached outside observer would not have deep access to lay understandings. In this sense, hermeneutic understanding is not simply the act of translating sociological into lay concepts. Rather, it is some degree of practical participation in members' lifeworlds and social practices that makes interpretation possible (Burgess 1982; Dreyfus 1980; Winch 1958). Understanding how members interpret the world and account for their actions involves more than just being able to understand the explicit verbal statements and reports that lay members may give, for example, in structured and standardized interviews (Becker and Greer 1970; Cicourel 1964; Schatzman and Strauss 1973). As opposed to mainstream survey research, the interpretive paradigm maintains that coding responses into data according to prefabricated rules of correspondence and operationalization is hardly sufficient to "really understand" members' definitions of reality.[5]

Therefore, hermeneutic understanding often involves the practical problem of being accepted as a (virtual and temporary) member of natural lifeworlds. In fact, some interpretive researchers (see M. Bloor 1983) regard being able to "get along" in natural lifeworlds as a sign of having understood correctly the meanings of social practices and lay interpretations. Understanding is validated by acceptance as a virtual member, and by being acknowledged by lay groups as someone who knows how to act in accordance with the group's rules:

> *Understanding* a symbolic utterance generally requires participation in processes of communication. Meanings, be they embodied in actions, institutions, products, words, patterns of cooperation, or documents, can only be decoded *from inside*. The symbolically prestructured reality is a universe that remains inaccessible and incomprehensible to an observer who is unable to communicate. The lifeworld is accessible only to a subject who makes use of his or her capacity to speak and act. He or she gains access by participating in the communications of members, and by becoming a virtual member of the group. (Habermas 1984:165)

I would say that this practical, participatory, and lay-oriented character of hermeneutic understanding is what distinguishes the interpretive paradigm most sharply from mainstream social research. Whereas mainstream research regards lay accounts primarily as data sources, believes that professional concepts and theories are generally superior to lay concepts and theories, and accepts institutions such as significance tests as validators of sociological knowledge, the interpretive paradigm maintains that a more realistic and less reified sociology requires *less* professional baggage and more direct and "unbiased" interaction with lay members (Filstead 1970a:2).[6]

Gaining knowledge through interacting and participating is different from following a set of methodological rules (Morgan 1983). Interacting relates to methodology like knowing how relates to knowing that, or like practical and tacit relates to explicit and algorithmic knowledge (Garfinkel 1967). Understanding a group from the inside is like gradually getting to know a person: both cannot be formalized and made predictable according to prior rules, and both involve "hermeneutic circles" more than deductive logic. In this sense, hermeneutics is a life-form, not a methodology.

To be sure, I believe a similar argument can be made for the positivist paradigm in social research. As a characteristic way of doing sociology, positivism is confident about the superiority of scientific method, believes in an external reality of objective facts, and values professional techniques of knowledge control over interpretive closeness and hermeneutic charity. The difference between hermeneutics and positivism is not that the former is a contingent life-form while the latter is a neutral methodology which helps us discover the truth about objective reality. The difference is that positivism expresses more confidence in objectivity, truth, and professional superiority. In this sense, positivism characterizes the self-understanding of fields that do Kuhnian normal science with more routinized and consensual practices. But these practices are still *social* practices, not the abstract rules of some philosophically authorized methodology. Hermeneutics does not relate to positivism as fiction relates to truth, or as stories relate to science, but like uncertainty and organizational openness relate to routinization and organizational closure.

In any case, I have chosen the interpretive paradigm to illustrate my point because practitioners of interpretation are somewhat more attuned to the notion that their practices are not and cannot be governed by rigidly standardized methodological canons. However, paradoxes do arise whenever *Verstehen* is misunderstood as methodology. Norman Denzin's (1978) symbolic interactionist methodology of "behaviorist naturalism" will serve to illustrate my argument. However, I would argue that his position and its paradoxes are typical of the interpretive paradigm insofar as it conceives of *Verstehen* as a methodology.

THE PARADOX OF INTERPRETIVE METHODOLOGY

From a symbolic interactionist perspective, Denzin (1978) argues that the "research act" itself is a process of symbolic interaction between lay subjects and professional researchers. Since the social world is structured symbolically through presociological lay interpretations, researchers must not arbitrarily impose their concepts and frames of reference on those of members. Instead, lay interpretations must be taken seriously as the *premises* of sociological accounts. To gain insight into natural lifeworlds, researchers must *participate* in episodes of symbolic interaction, for they cannot simply observe what is going on.

Since the research act itself is symbolic interaction, it is fundamentally subject to the same dynamics as ordinary interactions. That is, participants in interaction present and re-present self-concepts, define situations, negotiate the meanings of contexts, take each other's roles, and routinely account for settings. Consistent with the interactionist position, Denzin argues that these properties of interaction inevitably shape its outcomes. For the manifest content and dynamics of *any* interaction covary with its situational contexts, with participating interactants and their self-presentations, and with the idiosyncracies of particular encounters.

For example, in the sociological interview (Denzin 1978:112ff.), reality construction (i.e., the production of "data") is not neutral, objective, or context-independent but depends very significantly on the variable personal, situational, and contextual properties of the interaction between interviewer and respondent. The interview, Denzin argues, does not differ substantially from any other "normal" episode of interaction.[7] Most importantly, the interactive dynamics of interviews fundamentally shape the construction of social reality, such as the manufacture of responses as "data."

However, Denzin (1978) is also firmly committed to the research act as a general *methodology* of interpretation. That is, as a methodology, the research act must yield accurate descriptions and valid and reliable data. However, valid and reliable data do *not* depend on the contexts and dynamics of particular interaction episodes, for they are supposed to represent objective reality independent from contexts, situations, and individual interactants.

At this point, Denzin's position turns paradoxical. As a symbolic interactionist, he claims that there are no significant differences between interviews and other interactions. But as a methodologist, he insists on the interview generating valid and reliable data that represent objective reality, regardless of contexts and situations. From an interactionist perspective, the situational and contextual dynamics of interaction are inherent properties of all encounters, but from a methodological perspective, Denzin (1978:126) warns that "the interaction between an interviewer and subject may itself create sources of invalidity." Denzin first shows that research acts *are* interactions, but to obtain reliable and

valid data, he wants to eliminate or control for the very properties
of interaction as sources of invalidity. Instead of acknowledging
that once we accept that reality is a symbolic construction we
dismiss the idea of an objective and independent reality that can be
mirrored by valid and reliable descriptions, Denzin, being a good
positivist here, maintains this latter notion, and wants to purify the
research act from the "distortions" inherent in ordinary interaction. In fact, as a general methodology the interpretive paradigm
must paradoxically claim that there is such a thing as a privileged
interaction, namely the research act, that does not have the properties of an interaction, or that can "control for" these properties to
obtain valid and reliable data.[8]

I think we can avoid such paradoxes if we understand Verstehen as a way of doing sociology, or as a life-form, rather than as
a formal methodology. This does not exclude attempts at making
interpretation more "reliable" and "objective," but it does imply a
farewell to the grand old notions of Truth and Scientific Methodology. We may develop certain conventions and tools and tricks that
make Verstehen less idiosyncratic and subjective, just like positivist
techniques are a bit less idiosyncratic and subjective. But I don't
think such tricks will be any different from those we use when
trying to understand people. We still would have no idea of how to
arrive at "accurate representations" of reality by following a "scientific methodology" of interpretation. We would just be a bit
more certain that we were on the right track.

Interpretivism is more sympathetic to this notion than positivism. Interpretivists insist that nonmethodological and nonscientific
abilities are required for doing social and, especially, qualitative
research. Researchers must have "intuition" and "insight" (Taylor
1971), they must master the "art," not just the methods, of interviewing, and they must have certain "social talents" and be able to
establish rapport with their subjects and respondents (Warwick
and Lininger 1975; Rossi, Wright, and Anderson 1983). Similarly,
field research is not just a data-gathering device; rather, it is an
"adventure" with profound consequences for one's own style of
life (Glazer 1972). The researcher who wants to understand a
lifeworld from inside must to some degree change his or her own
life to become a virtual member of the group studied:

Thus, in the sciences of man insofar as they are hermeneutical there can be a valid response to 'I don't understand' which takes the form, not only 'develop your intuitions' but more radically 'change yourself'. (Taylor 1971:47f.)

This is why sourcebooks on field and qualitative research often caution their audience not to mistake informal guidelines and personal advice for general methodological rules (e.g., Wax 1971; Burgess 1982; Lofland 1976; Strauss 1987). If we want to understand the unfamiliar lifeworlds of gangs, hobos, or the very rich, we face a situation similar to that of an ethnographic observer encountering an alien society. At first, we do not know how to interact and interpret, and we even do not know exactly what it is that we need to learn. Understanding such an alien lifeworld is possible only by prolonged interaction and by step-by-step socialization into an unfamiliar way of doing things and seeing the world.

Consequently, like an ethnographer, the interpretive researcher might find it necessary to spend years interacting as a virtual member with the group under study (Freilich 1977). Although, of course, different degrees of involvement are possible that define different field roles for the researcher, the participant observer must develop intimate familiarity with his or her subjects (Gans 1982:54). Gradually arriving at a deep understanding of a lifeworld from within, interpretive researchers might then attempt to translate their experiences into more professional and "scientific" concepts and theories that go beyond the symbolic universe of the group under study. Or, as phenomenologically inclined interpretivists say, we must gradually replace members' "natural" and more or less "reified" attitudes by the "theoretical stance" of the professional researcher.

But the point of the interpretive paradigm is that we must first understand social worlds from within, before we develop scientific models and explanations. These models and explanations have not privileged but parasitic status because they are and remain dependent on lay accounts. That is, we must inductively "ground" our scientific theories in the everyday knowledge of lay members, and we must not impose our professional frames of reference on members' frames. More radical versions of the interpretive paradigm even claim that translating members' accounts into scientific accounts is pointless, for such a translation would not yield any cognitive gains, but would simply replace one language game by another (Winch 1958).

HERMENEUTICS AS ORGANIZATIONAL STRUCTURE

If hermeneutic understanding is not a methodology, then two important issues remain to be addressed. First, given that the modern sciences are established as more or less autonomous work organizations, what precisely does it mean to practice hermeneutics as a life-form? And second, is hermeneutics the only life-form available in the social sciences, and if this is not so, what determines our choice between alternative life-forms?

Richard Rorty (1979) goes a long way toward answering the first question. He concludes his discussion of hermeneutics arguing that accepting lay definitions of social reality as the premise of sociological accounts is based on a *moral*, not an epistemological or ontological, decision. What Rorty means is that it is not social reality itself that forces sociology into accepting lay interpretations as the grounds for sociological interpretations. We *can* ignore lay knowledge, because it is not reality itself that selects its own accurate descriptions, but communities of knowledge producers. And these communities may decide to take lay interpretations very seriously, like the interpretive paradigm, or to accept them as data sources only, like mainstream social research, or they may decide to ignore lay interpretations altogether, like classical behaviorism.

Rorty does *not* say, and I agree, that there are no excellent reasons to do interpretive (or, for that matter, positivist) research. From a pragmatist view, I would say that interpretivism and positivism are useful ways of answering different kinds of questions. If we want to deeply understand the ways in which people live their lives and experience their worlds, we should probably do an interpretive case study. This is what Habermas (1970) calls an "interest in maintaining mutual understanding," which he believes underlies the hermeneutic disciplines. But if we are more interested in predicting what large numbers of people will do and in controlling their behaviors, then positivism is probably more adequate.

Habermas calls this the "interest in instrumental control" underlying the "empirical-analytic" sciences. In any case, I think the epistemological notion of "reveals the true nature of reality" should give way to the pragmatist notion of "serving a particular purpose well or badly."

According to Rorty (1979), what makes the decision to accept members' definitions of reality a moral one is our commitment to the Western humanistic tradition of social and philosophical thought which celebrates individuals as capable and knowledgeable agents who can be expected to be responsible and accountable for their own actions. For Rorty (1979), hermeneutics expresses our respect for this tradition of praising and worshipping the autonomous modern self. Hermeneutics insists that agents themselves know best what they are doing and why they are doing it. Hence, the interpretive paradigm feels a *moral*, not ontological or epistemological, pressure to accept lay definitions of social reality as the starting point of its own descriptions. The *epistemic* relation between professional and lay accounts turns into a *social* or moral relation between groups of people. This is "sociological philosophy." The point of Rorty's (1979:349) argument is that as a moral decision, hermeneutics does not react to the pressures of reality, but to the pressures of morality:

> The familiar claim that a speaker's description of himself usually needs to be taken into account in determining what action he is performing is sound enough. But that description may perfectly well be set aside. The privilege attached to it is moral, rather than epistemic. The difference between his description and ours may mean, for example, that he should not be tried under our laws. It does not mean that he cannot be explained by our science.

My only quarrel with Rorty is that by interpreting hermeneutics as a moral decision, he disregards the *structural* arrangements of scientific communities. I do agree with Rorty that there is nothing in the structure of social reality that forced sociology into hermeneutics, but I would say that hermeneutics involves more than moral decisions. Since the sciences are more or less professionalized and collective systems of knowledge production, moral decisions are embedded in the social *structures* of scientific organizations.

That is, I want to suggest that hermeneutics, or the acceptance of lay interpretations as the premises of sociological accounts, corresponds to a particular structure of the sociological profession as a reputational work organization. In this context, I shall concentrate on two forms of organizational control: internal and external control.

External Control. "External control" refers to a scientific organization's relationships to its environment of other scientific organizations, funding agencies, and the lay public. External control is primarily exercized through an organization's level of reputational autonomy. The level of reputational autonomy measures the extent to which a particular scientific organization is independent from other scientific organizations and the lay audience in setting the standards of knowledge production. A low level of reputational autonomy means that other scientific organizations and the lay public can, at least to a certain extent, influence the ways in which knowledge is manufactured. Conversely, high reputational autonomy indicates that a particular scientific organization exclusively controls the norms of scientific work, so that other disciplines and lay audiences cannot to a great extent influence a given field and its ways of producing knowledge.

I want to suggest that the decision to follow an interpretive approach to social reality is not a methodological decision that reality itself, by virtue of its symbolic nature, makes; nor simply a moral decision based on our respect for the modern self. The idea that lay interpretations should be taken into account is very useful and legitimate, but, as pragmatism has shown, the pressures of reality are never strong enough to force us into one type of representation only. If we decide to accept lay interpretations as the premises of our own accounts, we vote for an organization with comparatively weak external control. If we decide not to do so, we express our preference to live in a privileged and rather elitist organization with more self-referential and professionalized discursive practices. Both are decisions about ways of life or of doing sociology; neither is in any metaphysical sense "closer to the truth." Positivism means that we strongly believe in the superior rationality of scientific method, and that, therefore, professionals with credentials, training, and expertise are in a privileged cognitive position *vis-à-vis* lay people. Lay accounts are taken into account only as sources of information about the social world. As positivists, we decide to construct and validate knowledge according to professional and scientific criteria only. Lay accounts enter the equation only because access to objective facts is not immediately given, so that we must take a *detour* through lay responses as indicators of some deeper underlying reality. In short, positivism is, organizationally, the ideology of elite professionalism.

The interpretive paradigm, on the other hand, replaces the objectifying and distancing glance of positivism by a more communicative and participatory attitude. Interpretivism is skeptical of the privileged rationality of scientific method, and accepts lay members as competent cointerpreters of the social world. Lay meanings are not given and obvious as interview responses, but must be explained by lay members and decoded by *Verstehen* and prolonged interaction. Interpretive knowledge is gained not in the neutral perspective of the outside observer, but in the inside perspective of the participant. The construction of interpretive knowledge is, in a certain sense, parasitic upon lay interpretations, for the sociological researcher cannot simply assume to know more or even as much as regular members.

The actual local *construction* of interpretive knowledge, then, takes place in close interaction with lay members, not so much in the remote academic space of professional discourse. The interpretive researcher cannot easily claim cognitive privileges, for members are accepted as competent and knowledgeable subjects. In some versions, the interpretive paradigm grants members even the competence to *validate* knowledge claims (M. Bloor 1983).[9] Researchers cannot be sure whether their understanding is adequate until they are able to act like a group member would, and until the group accepts such understandings as accurate. More fully professionalized and autonomous scientific organizations monopolize validation standards, and count on controls such as "significance tests" that keep validation inside the professional community. But interpretivism deemphasizes sharp distinctions between lay and professional worlds and votes for a more lay-oriented and participatory social practice of knowledge.

Hence, it should not come as a surprise that advocates of the interpretive paradigm often describe themselves and their approach as "humanistic," "liberal," or even "radical" (Becker 1967; Schatzman and Strauss 1973; Bruyn 1966; Schwartz and

Jacobs 1979). The interpretive paradigm corresponds to a reputational organization that is more open to democratic and participatory epistemic practices. Positivism, on the other hand, is more likely to emphasize professional distinction and demarcation. The interpretive organization does not greatly emphasize exclusive membership criteria, or would declare these criteria—such as credentials or familiarity with the authoritative scriptures of a field— as not essential for *doing* social research. For example, in their effort to "ground" theory in lay concepts and definitions of reality, Glaser and Strauss (1967:27) recommend "literally to ignore the literature of theory and fact in the area under study, in order to assure that the emergence of categories will not be contaminated by concepts more suited to different areas." Due to its pronounced lay orientation, the interpretive life-form does not require "legitimate" contributions to sociological knowledge to be linked to credentials and professional membership.

To be sure, interpretivism, as an academic movement, remains situated within the institutions of collegiate control. The systems of peer inspection and evaluation continue to be relevant and temper lay control and participation in knowledge. In interpretive research, lay control and participation are most visible in the actual local and contextual *construction* of knowledge, not so much in the evaluation and professional certification of that knowledge. Naturally, construction and evaluation are not entirely independent, for interpretive researchers must keep an eye on anticipated collegiate responses to their work. But it does make a difference in the ways of doing sociology whether knowledge is constructed in the separate, autonomous, and privileged space of professionalized discourse, like in positivism, or in close interaction and communication with lay members. Even if professional colleagues remain important in rewarding interpretivist contributions to knowledge, the local and contextual *fabrication* of that knowledge occurs in more direct and often extended social contact with lay people, not in the "pure" realm of "theory construction," "operational definition," and "scientific method."[10]

Internal Control. "Internal control" refers to the ways in which the material and symbolic means of scientific production are distributed within scientific fields and groups. The theory of scientific organizations suggests that "task uncertainty" and "mutual dependence" are the two most important variables determining the structure of internal control. High uncertainty means that problems are not well understood, methods not clearly defined, and findings open to many diverse interpretations. Conversely, low uncertainty means that problems are clearly stated, methods prescribe straightforward research steps, and results are not too equivocal. Mutual dependence measures the extent to which scientists depend on peer groups and networks for obtaining crucial symbolic and material resources necessary for their work. High mutual dependence is typically found in fields where very expensive lab equipment is highly concentrated, and where closely coupled elite groups control the most prestigious means of scientific production. Conversely, low mutual dependence characterizes fields in which material and symbolic resources are more dispersed, and control is structurally fragmented between various independent organizational centers.

I want to suggest that in addition to transparent organizational boundaries, or weak external control, hermeneutics defines the overall ideology of fields with low mutual dependence and high task uncertainty, or weak internal control. This is the second, more general meaning of hermeneutics. In this general sense, hermeneutics does not refer to the lay-oriented practices of the interpretive subdivision in social research, but to the theoretical self-understanding of the discipline as a whole, which is coined by social theory and the philosophy of the social sciences.

Hermeneutic ways of self-understanding emerge in fragmented fields with high uncertainty. Under these conditions, the overall field is split into fairly independent practitioner networks, local research clusters, and regional and specialty associations with their separate publication outlets (Turner and Turner 1990). Organizational fragmentation supports more diversity in intellectual production, so that various "schools" and perspectives coexist without successful attempts at integration (Mullins 1973).[11] Each specialty and separate perspective develops its own characteristic rationales and research styles, and no school can claim, without strong opposition, that its ways of doing sociology are superior to all the competing alternatives.

Such structural fragmentation supports a pronounced paradigmatic pluralism (Ritzer 1980). As a result, task uncertainty increases, for there are no strong and persistent agreements between the various approaches. Each endorses its own way of practicing research, and so there is not a great deal of overall consensus even about basic issues ("What is social action?"), although agreement *within* schools may be more solid.[12] In this situation, there is not a lot of certainty about the right ways of explaining social reality, how to do research properly, and how to interpret the findings.

I want to suggest that fields with high internal fragmentation and task uncertainty engage in certain *types* of debates that are characteristically different from those in more integrated and routinized fields. I have called the debates in the former fields "hermeneutics" or "conversation," while the latter fields produce "science" and "facts". In conversational fields, intellectual exchanges, if they take place at all, are of a very informal and highly controversial sort. Due to the lack of stable and agreed-upon "normal" research practices, such fields worry a great deal about metaphysical and foundational issues. These ongoing philosophical debates create a strongly discursive and nonscientistic intellectual *habitus*. In sociology, prominent examples are the reconstructions of the presuppositions that underlie sociology's "theoretical logic" (Alexander 1982–83), the everlasting fact/value debates, the notoriously irresolvable micro/macro problem, or the *Methodenstreit*.

Since they lack stable and consensual *empirical* research practices, conversational fields are constantly inspecting their *transempirical* dimensions. They are not sure whether they can discover the objective *world*, and so they turn instead to the *word* and reflexively scrutinize their own epistemic practices (Ashmore 1989; Mulkay 1985). Conversational fields doubt whether they can or should be "scientific" and "objective," whether they do or do not have a unique methodology or any methodology at all, and they are skeptical about ever cumulating toward Truth.

Such philosophical skepticism and metatheoretical foundationalism contrasts with the more confident and self-assured pragmatism of more integrated and routinized fields (Kuhn 1970). Their cognitive styles are less philosophical and geared more toward solving concrete puzzles, "getting things done," and "making things work" (Knorr-Cetina 1981). As a result, there is more faith in science, objectivity, and cumulation.

Conversational fields, though, have not much faith in their future and in cumulation, and so they are strongly oriented toward their past instead. In sociology, there are constant interpretations and reinterpretations of the classics, and a busy subspecialty is dedicated to exegesis of the Holy Scriptures. In fragmented fields with multiple perspectives, the classics operate as a kind of "paradigm substitute" (Alexander 1987), which creates a certain, if minimal, sense of social solidarity and belonging to the same group. Of course, there is not much agreement on how to interpret the classics, either. In any case, conversational fields often merge their history with their systematic research, while more integrated and "normal" fields strictly separate history from systematics.

Since separate paradigmatic communities produce special rationales and ways of doing research, there is an overall sense of relativism in conversational fields. Knowledge comes to be seen more as an active and often subjective social construction rather than as a neutral representation of reality. There is a strong awareness of the cultural relativity of various *epistemes*, and they are viewed as contingent and selective choices from other possibilities (Luhmann 1984).

Due to low mutual dependence, there is some room and encouragement for highly individualistic and, sometimes, idiosyncratic ways of doing research. Conversational fields establish only loose and informal cognitive controls, so that practitioners have considerable discretion and autonomy in their work. As a result, texts often bear the mark of their individual authors and their writing styles. Considerable value is placed on "good writing." Conversational fields thus display a structural resemblance with literature, and so sociology has recently been likened to rhetoric and literary narrative (Brown 1987; Agger 1989).

Conversational and scientific fields, then, engage in different *kinds* of debates. In conversational fields, debates are much wider in scope, more subjectivistic in style, more metaphysical and philosophical in nature, and often revolve around foundational issues such as action and structure, micro and macro, fact and value, or lifeworld and system. Due to high fragmentation and uncertainty, debates are unlikely to be resolved since each special approach and separate subfield has vested interests in its own continuation. In conversational fields, there is usually not even a good notion of what *could* settle disputes, for such notions will also be controversial. In a certain sense, this situation resembles Kuhn's (1970) descriptions of "revolutionary" science.

In scientific fields with more cohesiveness and less uncertainty, however, debates are much more likely to focus on more limited and manageable problems. Grand metaphysical and foundational issues do not figure prominently in the everyday workings of science. Although there might be also a great deal of competition and conflict, there is at least a better notion of what would settle controversies. In normal science, the positions are not as incompatible as the worldviews and ideologies in conversational fields, for there is some paradigmatic background consensus that restricts the scope and defines the terms of debates.

SUMMARY AND CONCLUSION

I have argued that hermeneutics and interpretivism have not so much to do with the ontology of social worlds but with the ways in which social research is socially organized. I have not offered any philosophical rationales as to whether we should do hermeneutics or science. Neither have I suggested any arguments that would help us decide whether hermeneutics and conversation or science and methodology are closer to the true nature of society. What I can offer is some insight into the social origins of the hermeneutic problem. Hermeneutics arises in fields with transparent organizational boundaries and loose internal control. This argument implies that hermeneutics becomes a problem *whenever* scientific fields are structured in this way, not just when they happen to study social worlds. The theory of scientific organizations would argue that we could have something like a "mature science of society" without a hermeneutic problem if the discipline was more closely integrated and fully professionalized. This would imply, however, living in a more bureaucratic and rigid organization. The social science of economics seems to be a bit closer to that structure and, correspondingly, does not have much of a hermeneutic problem but wins Nobel prizes. In any case, the present fragmentation of the field cannot simply be overcome by appeals to good will and better insight. We will engage in hermeneutics and conversation rather than in science and fact production as long as the structure of our profession remains loosely organized. But, we might not want to change that, for the conversational life-form is more pleasant than a bureaucratic one.

NOTES

CHAPTER 1

1. Throughout this book, I shall use the term *realism* as a generic term for the received model of orthodox philosophy of science, which includes *positivism* and *empiricism*. I realize that there are many, and many important, differences among these various doctrines, but I think that they all share a fundamental belief in true scientific knowledge being ultimately due to the forces of Reason and Reality. It seems to me that most of the disagreements among the various realist epistemologies concern the ways and conditions under which representation is possible.

2. Mertonian sociology of science is not a unified paradigm. There exist a variety of approaches and there is considerable debate between various practitioners. But it is probably not unfair to say that as opposed to constructivist studies in the sociology of scientific knowledge, Mertonian researchers more or less believe that science has privileged cognitive status because its technical and social norms assure superior rationality and epistemic objectivity. Science is held to be different from religion and ideology, and there is some confidence in the self-regulating mechanisms of social control in science that temper deviance and nonconformity.

3. This is also true for those more recent Mertonian studies that Zuckerman (1988) describes as "structural" studies in the sociology of scientific knowledge. In Zuckerman's (1988:541–42) words, in these studies "rationality and rules of empirical evidence are regarded as primarily determining scientists' acceptance or rejection of truth claims, although, on occasion, nonrational social influences may reinforce that acceptance." Of course, this position reinstalls the tired old philosophical dualisms which the Strong Program in the sociology of scientific knowledge abandons.

4. There is a great deal of variability and debate in the field that makes classifications very difficult. Woolgar (1988a:14, n.1) remarks that "sociology of scientific knowledge" is a narrower term than "social studies of science," for the former comprises a body of studies that are "informed by historical and cultural relativism." For a recent review and attempt at delineating common themes, see Zuckerman (1988).

5. I agree with Restivo (1983), however, that the importance of Kuhn for the sociology of scientific knowledge has been exaggerated. Kuhn is more orthodox and less sociological than his followers realize but, undoubtedly, the *reception* of Kuhn's work has done much to develop constructivism.

6. The amount of diversity and controversy in the sociology of scientific knowledge is highlighted by the fact that there are scholars who would disagree

217

with either part of the previous sentence. Woolgar (1988a), for example, would say that there is no such thing as the "actual reality of science," and Latour (1987, 1988b) would contend that "social factors" are only one part of the networks that "technoscience" tries to enroll.

7. The exception is the network theory of technoscientific artifacts, see Latour (1987, 1988b); Bijker, Hughes and Pinch (1987); Callon, Law and Rip (1986).

8. What is also left out are the questions raised by "critical" sociologies of science, such as the emergence of modern science in the context of industrialization and bureaucratization, the alliances between science and the state, or the capitalist pressures on commodifying science as alienation; see Restivo (1988, 1989). These issues are even more macro than the ones raised by the theory of scientific organizations, and so will not be dealt with systematically in this book.

9. Again, the exception is the network theory of technoscientific artifacts that replaces the sociological explanation of science by an "associational" model of science in which "actor-networks" comprise heterogeneous alliances of social and nonsocial forces.

10. For a recent collection of essays in this more structural mode of constructivist inquiry, see Cozzens and Gieryn (1990). Structural constructivism is more theoretical, and there is a strong concern to bring the science field back to the core concerns of the larger discipline.

11. I shall discuss these issues more extensively in chapter 2.

12. See, for example, Latour (1987, 1988b); Bijker, Hughes, and Pinch (1987); and Callon, Law and Rip (1986).

13. However, I do think that the "social" parts of such networks are in a privileged position. Consider, for example, Latour's (1988b) marvelous analysis of the changes introduced into French society by the Pasteurian movement. The Pasteurians seized the hygienist movement, translated their practices, and redirected their way through Pasteur's laboratory where he became the sole spokesperson of the microbe (bacillus anthrax). Latour writes that "we cannot reduce the action of the microbe to a sociological explanation, since the action of the microbe redefined not only society but also nature and the whole caboodle" (p.38). However, the hygienists were a social movement, and it was a group of people, the Pasteurians in their laboratory, that defined what the microbe was and what it could do. I believe we should not easily give away the cognitive gains won by the social construction argument.

14. I shall make this argument in more detail in chapter 2.

15. I am thinking here of studies such as Bazerman's (1988) analysis of the emergence of the experimental report, or the various semiotic analyses of scientific texts discussed in chapter 3.

16. The point of this distinction is that it is very naive to assume that there are no important sociological differences between texts. Only when we continue the philosophical search for truth and adequate representation will we conclude that there are only texts and stories. In a philosophical sense, there are only stories, for we have no way of knowing which story is more-than-a-story because we have no way of knowing which story represents objective reality. But from a sociological perspective, there are important differences between fictional and scientific stories. Scientific stories can draw upon more powerful textual and nontextual resources; see chapter 3.

CHAPTER 2

1. The sociology of scientific knowledge is by no means a unified field. There are considerable differences in the areas of science studied, in the methods chosen for investigation, and in epistemological commitments. Therefore, it is probably more appropriate to define the field negatively: practitioners more or less unanimously reject the Mertonian institutional sociology of scientists and the realist philosophical model of science. In Zuckerman's (1988:546) words, the new program "calls into question the rationalist and objectivist accounts of science, which hold that logic and evidence are prime determinants of scientific validity and scientists' theory choice." With this caution in mind, I shall use the terms "sociology of scientific knowledge," "social studies of science," and "constructivism" interchangeably to distinguish the new program from the "Mertonian paradigm" or the "orthodox sociology of science," and from realist epistemology.

2. See, for example, the debate between Gieryn and advocates of the new program in *Social Studies of Science* 1982,12:279–335.

3. For recent reviews see Zuckerman (1988) and Woolgar (1988a).

4. The exception is the reflexive self-inspection in the sociology of scientific knowledge. Reflexivists such as Woolgar (1988b) and Ashmore (1989) would deny that there is such a thing as the "actual reality of science" that can be mirrored by "accurate representations." Hence, the sociology of scientific knowledge cannot claim any special epistemic status for its accounts, just as it denies such status for the accounts of science. Philosophical realism and constructivism cannot be distinguished on the grounds that the former misses actual scientific practice, while the latter gives us a true picture of science as it actually happens. Nevertheless, I doubt whether reflexivists really believe that Popper and Lakatos are as right or wrong about science as Latour or Collins.

5. Note the flagrant irony here: realist philosophy is rejected because it misses actual scientific practice in stating that natural science corresponds to actual reality, and constructivism replaces realism as a more adequate account of science as it actually is! Paradoxically, realist status is claimed for the constructivist claim that there is no such a thing as "realist status" for the accounts of science itself. It is this ironical paradox that reflexivists such as Woolgar (1988b) and Ashmore (1989) take as their starting point to develop New Literary Forms that avoid any realist connotations; see below.

6. Habermas (1984) gives an overview of the available candidates. The general trouble with "universal principles" and "context-independent standards" is that despite their alleged universality and transcendental status, they are not as undisputed and self-evident as they *should* be were they truly universal and foundational.

7. The fundamental *non sequitur* of relativist arguments is: to show that scientific accounts are "underdetermined" or even "undetermined" by nature does *not* imply that these accounts are not determined at all. The radical relativist credo of "anything goes" is sound *only* if we tacitly maintain the image of science as solely reacting to the pressures of reality because anything goes argues for the absence of these *epistemic* pressures only. Only if we continue to search for philosophical or transcendental justifications of knowledge will our inability to find them lead to the conclusion that there are no justifications whatever and that

anything goes. But to say that science does not react to epistemic pressures (of language, logic, and reality) is *not* to say that it does not react to any pressures at all.

For example, Andrew Pickering (1981) has shown that the controversy over the discovery of "magnetic monopoles" in the mid-seventies was structured by a limited number of socially available and acceptable interpretations and concepts. Pickering argues that in principle, the relevant evidence could have been interpreted in a variety of ways, but the actual debate revolved around only two conflicting accounts: magnetic monopoles versus fragmenting nuclei. That is, although *in principle* an infinite number of alternative interpretations might always be possible, in reality there are social constraints on what counts as "legitimate" science (see also Harvey 1981:106). Especially in noncontroversial or "normal" science fundamental paradigmatic dogmas and experimental routines considerably limit the possibilities of alternative interpretations (Kuhn 1970). It is these *social* pressures that relativist philosophies ignore. In this sense, they are no different from their realist predecessors.

That is, only from an abstract philosophical perspective can we say that alternative interpretations of the evidence are always *equally* possible. But in reality, some possible interpretations are always considered less plausible than others, some accounts are, as Canguilhem and Foucault say, not "in the truth," and some descriptions do not fulfill disciplinary criteria of "valid" and "legitimate" science. In other words, what is philosophically or logically possible may be socially absurd, that is, ruled out by traditions, conventions, and acceptable scientific practices. And what is socially possible in science can be expected to vary across time and disciplinary fields so that some fields—like sociology—are more relativistic and tolerant toward dissident views and approaches than others, such as physics, for example. In short, the number of "possible" alternative interpretations of some body of empirical evidence is not an abstract philosophical constant. The ahistorical philosophical notion of pure contingency or unlimited possibilities of alternative interpretations should be replaced by a sociological notion of varying degrees of "task uncertainty."

8. This is an oddly realist argument, for it grounds reflexivity in the reality of the subject matter in social studies of science. I shall argue that reflexivity has nothing to do with the area under investigation, but more with the social ways in which research is conducted. Besides, as Luhmann's (1989:22ff.) analyses of self-reference and autopoiesis show, reflexivity occurs whenever a semantic or communicative code is applied to itself. This may happen in science (Is it true to distinguish between true and false?), or in law (Is it legal to distinguish between legal and illegal?), or in any other social system.

9. Sal Restivo (1983) has argued that Kuhn's relativism is actually more of a myth. It appears that Kuhn is much less radical than is believed by those who celebrate him as an antirealist icon.

10. I shall discuss this program in chapter 3. The main point of Collins's studies of controversies in contemporary science is that there are no such things as independent experimental replications that could settle controversies through unambiguous interpretations of the empirical evidence. Rather, controversies are closed through social factors and informal negotiations. Collins has claimed that his findings have been replicated by his own and other studies of controversies, and it is this seeming paradox that Ashmore investigates.

11. The oddness of this argument is reinforced by the curious implication that something that is social is somehow less real. Collins should realize that the implications of being labelled and socially constructed as someone who is "mentally retarded," "criminal," or "insane" are enormously real for the persons who go to jail or the asylum.

12. Woolgar (1988c) would probably call me a "benign introspector" who favors the "relax and stop worrying" solution to the problem of reflexivity (Ashmore 1989:97). And indeed, I believe that once we have dismissed the idea of representation we can continue making statements about the world, just like legislators continued to make laws after the downfall of the natural law doctrine as the foundation of law. We simply need no foundations for our practices, and occidental culture will not collapse if we simply continue to do what we have been doing: say interesting things about the world without casting any transcendental anchors. "Reflexivity" then simply means that a sociology of scientific knowledge should include a sociology of sociology, which is the point of Bloor's (1976) symmetry postulate; see below.

13. If interpreted realistically, Ashmore's (1989) book, for example, is hardly more than a critical literature review. We learn from it what practitioners in the sociology of scientific knowledge have to say about relativism, reflexivity, and the paradoxes of self-referential discourse. A realistic reading finds the countless recursive loops and interpretive deconstructions rather annoying. Of course, Ashmore would say that a "realistic reading" is precisely what he tries to avoid, but what other options do we have once we have accepted the point that there is no such thing as representation? My point is that we would have learned much more from a sociological study of the sociology of scientific knowledge instead of remaining exclusively concerned with its texts. Ashmore wants to "celebrate" reflexivity, but he ends up celebrating nothing but the core group of practitioners in the field. In this sense, reflexivism is a Durkheimian ritual in which the group celebrates itself as being more reflexive than anyone else.

14. See my discussion of textuality in the following chapter.

15. I think this is the most important point of Latour's (1988a) critique of reflexivism. Latour contrasts "infrareflexivity" with "metareflexivity" and advocates the former as a strategy that rehabilitates the world ("known") by giving up the obsession with the word ("knower").

16. For example, as "New Frontiers" in the sociology of knowledge (Woolgar 1988b), or as the "story that the story of stories cannot be told (Ashmore 1989:69)." Ironically, this deconstructionist pathos echoes the realist pathos to have discovered the rational foundations for culture. Deconstructionists do not deconstruct culture or practice, but only *texts*. Deconstructionism is earth-shattering only for those who believe that the stability of the world really rests on any foundations.

17. See, for example, the Stanford expectation states research. There is a lot of faith in cumulation and scientific rigor here; see Wagner (1984).

CHAPTER 3

1. Latour himself would probably not understand his own position as a "theory," because he is generally skeptical of explanation and generalization. This

obscures the fact, however, that his own network model of science is itself cast in general terms, such as "translation" and "enrollment." Abandoning such general notions amounts to surrendering to the hypercomplexity of the empirical world.

2. This "nesting" of statements in other statements is by no means unique to science but constitutes the very structure of ordinary language. Whenever a particular statement is being made, its intelligibility and plausibility draw upon a host of other statements which are tacitly presupposed to make sense of and give credibility to the statement suggested. It is these tacit background assumptions (Schutz 1967), "contexts" (Garfinkel 1967), or "frames" of meaning (Goffman 1974) that modern microsociology holds responsible for the facticity of social order. Again, we see that social and natural facticity are constituted and reproduced by the same basic social mechanism: not questioning the obvious, and building the questionable upon the unproblematic.

3. Of course, levels of professionalization and disciplinary exclusion vary across time and scientific fields. This variation will be one of the main arguments of the theory of scientific organizations.

4. This is not to say that only professional colleagues can turn statements into facts. Social support is not restricted to scientific communities, and not even to "social" agents in the narrow sense of the term. The construction of techno-scientific artifacts shows that the science/technology distinctions are mistaken in that many and very diverse agents must be enrolled to create black boxes (Callon, Law, and Rip 1986; Bijker, Hughes, and Pinch 1987). My focus on professional communities is an analytic simplification but, I think, a justified one because professionalization does provide scientists with a special social authority not shared by those outside their networks.

5. Shapin and Schaffer (1985) show how Boyle and the experimental program used three related technologies to produce matters of fact. The *material* technology assured that the physical integrity of the air pump helped reality reveal itself; the *literary* technology of Boyle's writings created the appearance that readers were virtually witnessing experimental events; and the *social* technology involved boundary-work to demarcate the legitimate experimental community from Scholasticism and pure Hobbesian philosophy. All three technologies operated as "objectifying resources" (p.77) in that the role of agency in constructing facts withdrew behind the lines of impersonality drawn by machines, texts, and social solidarities.

6. This is the same strategy as that employed by the Pasteurians (Latour 1988b). The Pasteurians tried to convince the hygienist movement that they had to pass through Pasteur's laboratory if they wanted to successfully fight epidemics. This process is like setting up a very strong market position or even a monopoly: to call long distance, one has to call AT & T *first*.

7. Latour (1988b) calls these "trials of strength." Trials of strength test the firmness of support networks. Reality and facts are whatever resist these trials.

8. More recently, this approach has also been applied to the social construction of technological systems; see Pinch and Bijker (1987).

9. But see Shapin and Schaffer's (1985) account of the controversy between Boyle and Hobbes over the experimental program. From their Wittgensteinian and Kuhnian perspective, Shapin and Schaffer interpret this debate as one over incompatible forms of life.

10. Reflexivists are quick to notice the ironical paradox in EPOR's claim to

have replicated the finding that there is no such thing as replication; see the previous chapter, "The Issues of Relativism and Reflexivity."

11. This, however, is a matter of degree. Science is always competitive and conflictual, only the means of conflict and competition change.

12. One of the main arguments of EPOR is that replications, the realist cornerstone of scientific objectivity, can never decisively settle controversies because they are controversial themselves (Collins 1985). Most importantly, there are always debates over what counts as a proper replication, who counts as a competent experimenter, and over the alternative explanations that can be offered to account for failures to replicate; see also Shapin and Schaffer (1985:229–30).

13. Again, the exception is the structural brand of constructivism arguing in the materialist and conflict tradition. For example, Collins and Restivo (1983b) have shown how the patterns of competition and conflict in mathematics have changed in response to the increasing academic professionalization of the field: "saintly intellectual politicians" replace ruthless "robber barons" once the community is organized in more collective and less patrimonial ways. The "ethos" of science is not universal, but coemerges with the advent of community organization in science.

14. I am using this term here broadly to include any study of the rhetorical and stylistic devices employed in scientific texts.

15. Trust is especially problematic once copresence has been replaced by community organization in science. Bazerman (1988) has followed the rhetorical and stylistic changes experimental reports underwent once experiments were no longer witnessed by copresent scientists. In order to convince distant peers, experimental reports started to employ a detailed language of representation that allowed for the virtual witnessing of experiments. The less scientists could rely on copresent colleagues to validate their observations, the more they had to employ the language of facts and evidence to convince their readers: "Since neither the reader nor any surrogates or representatives, except for the author himself, has witnessed the series of experiments, the account must stand in the place of the witness. The reader in order to understand the experimental argument must vicariously witness the experiment through the account. In order to earn the trust of the reader, the story of the experiments must be told plausibly if not persuasively, and the events reported on must provide sufficiently good cause for the investigator to come to the conclusions he reports (Bazerman 1988:74)."

Shapin and Schaffer (1985:60) make a similar point in their analysis of Boyle's "literary technology": "The technology of virtual witnessing involves the production in a *reader's* mind of such an image of an experimental scene as obviates the necessity for either direct witness or replication."

16. But see Woolgar (1982) for a critical comment on the "realism" of current lab ethnographies.

17. The most notable exception is Lynch's (1985) study that recommends the highest possible degree of familiarity with the technical intricacies of a given field.

18. This is probably the reason for the obsession of experimentalists with getting the material equipment to work, and with building new detectors which make possible new discoveries; see Traweek (1988:46–73). Also, the naiveté of deconstructionists becomes evident, for they deconstruct only texts, and are therefore rather harmless.

19. But highly stratified fields with high resource concentration do separate core from peripheral areas, and hence stratify access to equipment and work opportunities.

20. Possibly, industrial science is more likely to establish such authoritarian controls.

CHAPTER 4

1. Of course, as the field matures, debunking realism will become less of a routine exercise. But still, the field remains fascinated by philosophical problems, as the most recent debates on reflexivity and representation show. See chapter 2.

2. Or, case studies create hypervariability without a common metric that would make comparisons possible.

3. Two notable exceptions, Latour and Woolgar's (1986) "cycles of credit" and Knorr-Cetina's (1982) "transepistemic arenas" of research do not really relate community organization to the *contents* of science, but add this dimension onto microreconstructions of the research process.

4. See my analysis of fact production in the previous chapter.

5. Ever since Kuhn's (1970) attempt to identify the boundaries of scientific communities, there has been a great deal of controversy over what the appropriate units of analysis are for studying scientific "fields," "research specialties," "networks," etc; see Chubin (1976) and Geison (1981) for reviews. I believe the confusion over drawing the proper boundaries around units of analyses is partly due to the fact that the definition and maintenance of boundaries is a self-referential accomplishment of scientific communities. It is difficult for observers to identify the boundaries of communities because these communities themselves constantly define and redefine these boundaries. However, I believe the identification of exact boundaries is of secondary importance, and is not required for a general theory of scientific organizations (Collins 1988). For heuristic purposes, I shall adopt Whitley's (1984) definition of scientific fields as organizations that control scientific training, certification and credentials, the material resources needed for scientific production, and communication systems.

6. I am referring here to the well-known divisions in sociology between various theoretical approaches and research traditions; such as "mainstream" sociology, symbolic interactionism, ethnomethodology, "Chicano" sociology, etc.

7. I have borrowed this term from Rorty (1979), who himself borrowed it from Oakeshott.

8. I am aware of this classification being highly schematic. Not all of the natural sciences have the same organizational structures and cognitive styles, just as the social sciences differ on these variables. Hence, I would also expect variations between various natural sciences, and between various social sciences; see Fuchs and Turner (1986), and Chapter 8.

9. I realize that post-Kuhnian epistemology has undermined realism as the philosophy of the natural sciences. But even postempiricists like Taylor (1971), Dreyfus (1980), Habermas (1984), Giddens (1984), or Rorty (1979) still maintain that there are some critical differences between paradigmatically integrated or mature and paradigmatically fragmented fields. And these differences largely correspond to those I shall describe shortly.

10. A more extensive and elaborate analysis will be given in chapter 7.

CHAPTER 5

1. The concept of a "match" between technology and structure has been subject to justified criticisms. It is not clear what exactly "match" means, or how it is brought about in organizations. I shall interpret "match" liberally to mean that work conditions impose certain restrictions on the possible variations in organizational structures. That is, I do not advocate a "deterministic" notion in which technology completely shapes structure.

2. I believe these imprecisions are partly responsible for the inconclusive empirical evidence produced by replications of Woodward's original study; see Pugh, Hickson, Hinings, and Turner (1968, 1969); Lincoln, Hannada, and McBride (1986); Blau, McHugh, McKinley, and Tracy (1976).

3. See figures 4.1 and 4.2, pp. 87 and 93. The only difference between my and Perrow's models is the degree of mutual dependence. But, as I will note soon, scientific organizations are generally based on *reciprocal* interdependence (or "coordination through feedback"), and hence in my model of scientific organizations, the degrees of dependence vary only *within* this type of interdependence. This point will become clearer in my discussion of Thompson's technological interdependence typology; cf. below.

4. Since I will be focusing on scientific organizations that are generally based on *reciprocal* interdependence (in Perrow's terms: "complex interactions"), I am concerned less with different types than with different degrees of interdependence.

5. See, for example, the Aston studies; Pugh, Hickson, Hinings, and Turner (1968, 1969). Most importantly, the Aston studies conclude that technology determines structure more in smaller than in larger organizations, because small organizations are entirely centered around the workflow. Scientific organizations are rather small in size, so we would expect strong technological effects on structure.

6. For critical comments on this measurement strategy see Duncan (1972), Tosi, Aldag, and Storey (1973); and Downey, Hellriegel, and Slocum (1975).

7. Lawrence and Lorsch (1967) also observed differences in goals, time orientations, and interpersonal attitudes between members of different departments. These differences also appear to be related to differences in the environments faced by various departments.

8. Shrum and Morris (1990) find that task uncertainty is also an important determinant in the interorganizational networks or sets that create complex technical systems.

9. See, for example, Emery and Trist's (1965) fourfold typology, Thompson's (1967) classification schema, or Weick's (1979) model of "enactment."

10. But do note the strong resemblances between the model and Perrow's (1967:199; 1984:97) typologies of organizational structures.

11. Therefore, the empirical focus of institutionalism on educational organizations is probably not accidental. For educational institutions combine fairly complex technologies with loose coupling of subunits and departments (Perrow 1984). Under these conditions, we would expect rationality to be more of a myth than a reality.

12. To the extent that institutionalist theory holds that rationality is a myth, it envisions the reality of organizations much in terms of garbage-can theory.

CHAPTER 6

1. I shall discuss this point in the following chapter.

2. As measured by the citations received by scientists' works.

3. The situation is different, it seems, for the scientific ultraelite of Nobel Prize winners. Zuckerman (1977) reports that "future members of the scientific elite were far more likely to go to elite colleges than the run of students in their age cohorts (p.86)," that "all seventy-four laureates (chosen between 1901 and 1972) with doctoral degrees from American universities were educated at only twenty-one [elite, S.F.] universities (p.88)," and that "more than half of the ninety-two laureates who did their prize-winning research in the United States by 1972 had worked either as students, postdoctorates, or junior collaborators under older Nobel laureates (pp.99f.)."

4. Zuckerman (1988) still spends twelve of forty-seven pages of her review of the science field on norms and scientific ethos, although the question has moved to the periphery of the field.

5. Collegiate professions are those in which professional practitioners define reality; patronage occupations are those in which clients and customers define their own needs and strategies.

6. This point will be discussed more extensively under the heading "mutual dependence."

7. Cole and Cole's (1973:122) main conclusion in their study of the scientific stratification system in physics is that "the quality of a physicist's work, as evaluated by his colleagues, is the single most important determinant of whether he rises to a position of eminence or remains obscure." "Quality" is measured by the number of citations a scientist's work receives.

8. See the more extensive discussion in chapter 4, "Mundane and Scientific Knowledge."

9. I shall discuss this point in more detail when comparing science to art, see below.

10. These seem to be those professionals who work in areas of high task uncertainty and innovation, cf. above.

11. Of course, these conventional distinctions retain their importance for education, professional socialization, occupational identification, and the like.

12. Since medical specialists are a more prestigious group than generalists, however, it will be more difficult to become a member of specialist communities.

13. This is why hospitals are often criticized for their "anonymity" and "factory-like" character. Hospitals are in a constant legitimation crisis because for them it is more difficult to claim that their main concern is the patient's health.

14. Some social sciences and humanities are less professionalized since they do not require heavily concentrated and expensive research equipment. One can still find the hobby historian and lay sociologist.

15. Hence, amateurs are forced into "marginal" and "peripheral" fields, such as parapsychology (Collins and Pinch 1982).

16. See chapter 3.

17. In some scientific fields, however, the customers of scientific research have more control over scientific production, as in much of industrial science and biomedical research; see Whitley (1984).

18. Correspondingly, we would expect collegiate control structures in theoretical fields to be generally more individualistic and patrimonial than in experimental and more quantitative areas.

19. This raises the question of whether literature really is a profession. Most standard treatments of the professions omit literature. But definitional problems are less crucial here than structural differences between work activities.

20. The situation is very different for writers of television programs, for example. The production of TV programs is much more concentrated and requires more extensive resources. Also, TV programs are scheduled to air in advance and at regular times so that writing them must be organized more systematically and predictably. This is why they are all alike.

21. Again, the situation is very different for writers of TV programs, who are often unionized and have considerable bargaining power.

22. Howard Becker (1982) has stressed the collective nature of artistic production. Against our traditional images of the creative artist producing art in lonely acts of imagination, Becker correctly argues that many people are required to complete and distribute a piece of art. Unfortunately, however, Becker lacks a *theory* of artistic production, and hence cannot explain *variations* in the social production of various art forms. The degree to which artistic production is a collective process involving high mutual dependence and collegiate control depends on the technologies and material resources required to produce art. Compare, for example, the movies and literature: due to high resource concentration and fairly expensive equipment, making movies is a much more "cooperative" and socially dense process than writing poetry or novels.

23. Characteristically, much theoretical work in sociology, such as exegetic interpretations of classical texts, has the same structure. We shall analyze this striking resemblance between social theory and art more extensively below.

24. As is most obvious in the aesthetics of Critical Theory (Adorno 1984).

25. The "general audience," however, is a very recent invention of modern mass markets for art and literature. In premodern times, aesthetic production and consumption was much more socially restricted. Hauser (1982:134ff.) has shown how art production and consumption developed from the medieval monasteries to the early modern mason's lodges and guild workshops, before becoming more individualistic in the Renaissance studios. The individual artist producing for an anonymous "general audience" is a very recent situation. It seems likely that the restriction of aesthetic production and consumption to religious, aristocratic, and bourgeois elites—patronage systems of one kind or another—greatly unified and objectified aesthetic standards. Even today, patronage continues to be an important source of support for many artists, but has generally been replaced by the more anonymous and autonomous dealer-critic system (Wolff 1981:26–48).

26. See Chapter II, pp. 60ff.

27. Of course, the term "postmodern" is very ambiguous and has many meanings. In the present context, I shall use the term "postmodern" to identify approaches in social theory that are skeptical of "science," and which liken science and sociology to literature.

28. I would say that the structural reason for this belief is that intellectuals spend a great deal of their time reading and writing *texts*, and hence come to believe that texts are all there is.

29. Rorty (1979), for example, makes this argument in what I would say is the best postpositivist statement available.

30. This is why David Wagner (1984) restricts the possibility of cumulation in theory to fairly narrow research programs, such as status expectation studies.
31. Collins and Restivo (1983b) make a similar point for the case of mathematics.

CHAPTER 7

1. A fact Whitley (1984:266ff.) himself recognizes.
2. This is not to say that certain subspecialties cannot be organized densely enough to produce "cumulation." But cumulation is then restricted to such subfields, which is why David Wagner (1984) limits the possibility for cumulative growth in sociology to such subspecialties.
3. See the following chapter.
4. Of course, cumulation and specialization need not be mutually exclusive forms of scientific change. Some fields will combine cumulation in uncertain core areas with specialization in more peripheral areas.

CHAPTER 8

1. By "life-form" I mean particular social practices, ways of doing research, and the forms of interaction implied by them. More specifically, in the present context "life-form" indicates the everyday ways of doing sociology and social research. For example, the life-form of an armchair social theorist differs considerably from that of a participant observer or experimental physicist. These groups of scientists inhabit different worlds, they talk to different kinds of people, and they do their work in characteristically different ways.
2. Various versions of the interpretive paradigm differ in the extent to which they accept members' interpretations as premises for sociological accounts. Options range from viewing lay interpretations as mere starting points for sociological analysis (Blumer 1969), to granting members the competence to validate sociological accounts. But all versions of the interpretive paradigm believe that lay interpretations constitute social worlds, that reality is a social construct, and that we must take members' accounts very seriously in our own professional work.
3. I should say again that labels such as "interpretive paradigm" or "positivism" cover very diverse approaches. Oral history is different from participant observation and unstructured interviews. But it seems fair to say that these approaches share a belief in the importance of letting lay people *speak for themselves*, rather than translating their responses *a priori* into the terms of professional discourse.
4. "Grounded Theory" (Glaser and Strauss 1967), for example, advocates an inductive strategy for theory building that starts out with and remains controlled by lay accounts. The movement between professional and lay accounts in interpretivism is circular, while that movement in positivist mainstream sociology privileges professional accounts and facts.
5. Thus, the main rationale of interpretive researchers for preferring unstructured over structured interviews, for example, is that the former resemble more

closely the patterns of ordinary and natural interaction between lay members (e.g., Warwick and Lininger 1975:132ff.; Kidder and Judd 1986:273ff.).

6. The opposition between mainstream and interpretive approaches is somewhat mitigated by the practice of "triangulating" various methods; see Bryman (1984). I have emphasized opposition rather than similarity for the *analytic* purpose of clarifying the practical implications of a dualistic (hermeneutic) versus a monistic (positivist) approach to social reality.

7. Of course, Denzin does not deny that ordinary and research interactions differ in that they pursue different goals. But these two types of interaction share the symbolic properties and dynamics of all interactions, even if the mundane differ from the research objectives.

8. Ironically, Denzin's position is very similar to the old and discarded logical positivist project to purify ordinary language from its intrinsic "ambiguities" and "inconsistencies." Wittgenstein said that these properties are not flaws to be repaired but constitute the very structure of everyday communication. His recommended therapy was not to scratch where it doesn't itch, which means, in our context, not to treat the very properties of interaction as sources of bias that need to be controlled.

9. It seems to me that such a decision, albeit radical, is difficult to avoid if safe cognitive privileges are denied to professional researchers.

10. To the extent to which the interpretive paradigm remains within a more or less professionalized scientific organization, it will likely claim some sort of privileged knowledge that is not easily gained by lay members. This explains why different versions of the interpretive paradigm grant lay members varying degrees of control over sociological knowledge. The more control members are granted, the lower the degree of independent professionalism in the discipline. Conversely, if lay members are granted some but low control over sociological knowledge, a certain amount of autonomy and professionalism is preserved for the discipline.

If a certain degree of reputational and professional autonomy is to be maintained, then the interpretive paradigm must somehow restrict members' knowledgeability. It is then assumed, for example, that "unintended consequences of action" systematically escape actors' accounts (Giddens 1984), or that "deep" and "latent" structures of social practices are beyond conscious control by members (Garfinkel 1967). Alternatively, it may be claimed that hermeneutics must be supplemented by some version of functionalism for the analysis of those macrostructures and macroprocesses that structure members' lifeworlds "behind their backs" (Habermas 1987).

Or, somewhat less plausibly, it is claimed that scientific concepts are "more abstract" and "more clearly defined" than common-sense concepts and that, therefore, sociological discourse is in some way superior to lay discourse (Schutz 1967; Denzin 1978). Thus, various analytical strategies employed by the interpretive paradigm restrict members' knowledgeability and aim at maintaining a certain amount of sociological professionalism. What escapes members' knowledgeability is then reserved as the exclusive domain of professional researchers. But insofar as the interpretive paradigm grants lay members some control over the standards of knowledge production, it reduces the degree of autonomous professionalism in sociology.

11. This is why attempts at integrating the field through cognitive mechanisms only are bound to fail. Gibbs (1989), for example, laments the fragmented

state of sociology, but his recommendation is only that all of sociology should adopt the "central notion" of "control." This proposal is sociologically naive, for it does not see that cognitive diversity rests upon and is perpetuated by organizational diversity.

12. But even in areas that count as rather "mature," such as formal organizations, there is considerable disagreement and debate. Population ecologists, resource dependence theorists, contingency theorists, institutionalists, and neoclassicists would disagree on even such a fundamental problem as to how to define what an organization is; see Scott (1981).

REFERENCES

Abel, Theodore. 1953. "The Operation Called Verstehen." Pp.677–87 in Herbert Feigl and May Brodbeck, eds., *Readings in the Philosophy of Science*. New York: Appleton.

Adorno, Theodor W. 1984. *Aesthetic Theory*. London: Routledge and Kegan Paul.

Agger, Ben. 1989. *Socio(onto)logy: A Disciplinary Reading*. Urbana and Chicago: University of Illinois Press.

Aldrich, Howard, and Sergio Mindlin. 1978. "Uncertainty and Dependence: Two Perspectives on Environment." Pp.149–70 in Lucien Karpik, ed., *Organizations and Environment. Theory, Issues, and Reality*. Beverly Hills: Sage.

Alexander, Jeffrey C. 1982–83. *Theoretical Logic in Sociology*. 4 vols. Berkeley: University of California Press.

Alexander, Jeffrey C. 1987. "The Centrality of the Classics." Pp.11–57 in Anthony Giddens and Jonathan H. Turner, eds., *Social Theory Today*. Cambridge: Polity Press.

Apel, Karl-Otto. 1981. *Transformation der Philosophie*. Frankfurt, Germany: Suhrkamp.

Aronowitz, Stanley. 1988. *Science as Power: Discourse and Ideology in Modern Society*. Minneapolis: University of Minnesota Press.

Ashmore, Malcolm. 1989. *The Reflexive Thesis: Wrighting Sociology of Scientific Knowledge*. Chicago: University of Chicago Press.

Austin, John L. 1962. *How to do Things with Words*. Oxford: Oxford University Press.

Barnes, Barry. 1974. *Scientific Knowledge and Sociological Theory*. London: Routledge and Kegan Paul.

Bazerman, Charles. 1988. *Shaping Written Knowledge: The Genre and Activity of the Experimental Article in Science*. Madison, WI: University of Wisconsin Press.

Becker, Howard S. 1967. "Whose Side Are We On?" *Social Problems* 14:239–49.

Becker, Howard S. 1982. *Art Worlds*. Berkeley: University of California Press.

Becker, Howard S., and Blanche Greer. 1970. "Participant Observation and Interviewing: A Comparison." Pp.133–42 in William J. Filstead, ed., *Qualitative Methodology: Firsthand Involvement with the Social World*. Chicago: Markham.

Ben-David, Joseph, and Randall Collins. 1966. "Social Factors in the Origins of a New Science: The Case of Psychology." *American Sociological Review* 31:451–65.

Bergesen, Albert. 1988. *The Ritual Order*. Unpublished Manuscript: University of Arizona.

Bernstein, Basil. 1974–77. *Class, Codes, and Control*. 4 vols. 2nd ed. London: Routledge and Kegan Paul.

Bijker, Wiebe E., Thomas P. Hughes, and Trevor J. Pinch, eds. 1987. *The Social Construction of Technological Systems: New Directions in the Sociology and History of Technology*. Cambridge: MIT Press.

Blau, Peter M., Cecilia McHugh, William McKinley, and Phelps K. Tracy. 1976. "Technology and Organization in Manufacturing." *Administrative Science Quarterly* 21:20–40.

Bloor, David. 1976. *Knowledge and Social Imagery*. London: Routledge and Kegan Paul.

Bloor, David. 1983. *Wittgenstein: A Social Theory of Knowledge*. New York: Columbia University Press.

Bloor, Michael J. 1983. "Notes on Member Validation." Pp.156–72 in Robert Emerson, ed., *Contemporary Field Research: A Collection of Readings*. Boston: Little, Brown and Co.

Blumer, Herbert. 1954. "What is Wrong with Social Theory?" *American Journal of Sociology* 49:707–10.

Blumer, Herbert. 1969. *Symbolic Interactionism: Perspective and Method*. Englewood Cliffs: Prentice-Hall.

Bogdan, Robert, and Steven J. Taylor. 1975. *Introduction to Qualitative Research Methods: A Phenomenological Approach to the Social Sciences*. New York: Wiley.

Bourdieu, Pierre. 1975. "The Specificity of Scientific Fields and the Social Conditions for the Progress of Reason." *Social Science Information* 14:19–47.

Brown, James, ed. 1984. *Scientific Rationality: The Sociological Turn*. Dordrecht: Reidel.

Brown, Richard H. 1987. *Society as Text*. Chicago: University of Chicago Press.

Bruyn, Severine T. 1966. *The Human Perspective in Sociology*. New York: Prentice Hall.

Bryman, Alan. 1984. "The Debate about Quantitative and Qualitative Research: A Question of Method or Epistemology?" *The British Journal of Sociology* 35:75–92.

Burgess, Robert G., ed. 1982. *Field Research. A Sourcebook and Field Manual*. London: Allen and Unwin.

Burns, Tom, and G. M. Stalker. 1961. *The Management of Innovation*. London: Tavistock.

Calhoun, Craig. 1988. "Sociology, Other Disciplines, and the Project of a General Understanding of Social Life." Unpublished Manuscript: University of North Carolina.

Callon, Michel. 1986. "The Sociology of an Actor-Network: The Case of the Electric Vehicle." Pp.19–34 in Michel Callon, John Law, and Arie Rip, eds., *Mapping the Dynamics of Science and Technology*. London: Macmillan.

Callon, Michel. 1987. "Society in the Making: The Study of Technology as a Tool for Sociological Analysis." Pp. 83–103 in Wiebe E. Bijker, Thomas P. Hughes, and Trevor Pinch, eds., *The Social Construction of Technological Systems*. Cambridge: MIT Press.

Callon, Michel, John Law, and Arie Rip, eds. 1986. *Mapping the Dynamics of Science and Technology: Sociology of Science in the Real World*. London: Macmillan.

Chubin, Daryl E. 1976. "The Conceptualization of Scientific Specialties," *The Sociological Quarterly* 17:132–47.

Chubin, Daryl E. 1983. *Sociology of Sciences: An Annotated Bibliography on Invisible Colleges*. New York: Garland.

Cicourel, Aaron V. 1964. *Method and Measurement in Sociology*. New York: Free Press.

Cole, Jonathan R., and Stephen Cole. 1973. *Social Stratification in Science*. Chicago: University of Chicago Press.

Cole, Stephen, Leonard Rubin, and Jonathan R. Cole. 1978. *Peer Review in the National Science Foundation: Phase One of a Study*. Washington: The Academy.

Collins, Harry M. 1975. "The Seven Sexes: A Study in the Sociology of a Phenomenon." *Sociology* 9:205–24.

Collins, Harry M. 1981a. "What is TRASP? The Radical Programme as a Methodological Imperative." *Philosophy of the Social Sciences* 11:215–24.

Collins, Harry M. 1981b. "Stages in the Empirical Program of Relativism." *Social Studies of Science* 11:3–10.

Collins, Harry M. 1981c. "The Place of the 'Core Set' in Modern Science: Social Contingency with Methodological Propriety in Science." *History of Science* 19:6–19.

Collins, Harry M. 1981d. "Son of Seven Sexes: The Social Destruction of a Physical Phenomenon." *Social Studies of Science* 11:33–62.

Collins, Harry M. 1983. "An Empirical Relativist Programme in the Sociology of Scientific Knowledge." Pp. 85–113 in Karin D. Knorr-Cetina, and Michael Mulkay, eds., *Science Observed: Perspectives on the Social Study of Science*. London: Sage.

Collins, Harry M. 1985. *Changing Order: Replication and Induction in Scientific Practice*. London: Sage.

Collins, Harry M., and Trevor J. Pinch. 1982. *Frames of Meaning: The Social Construction of Extraordinary Science*. London: Routledge and Kegan Paul.

Collins, Randall. 1975. *Conflict Sociology: Toward an Explanatory Science*. New York: Academic Press.

Collins, Randall. 1988. *Theoretical Sociology*. San Diego: HBJ.

Collins, Randall. 1989. "Toward a Theory of Intellectual Change: The Social Causes of Philosophies." *Science, Technology, and Human Values* 14:107–40.

Collins, Randall, and Sal Restivo. 1983a. "Development, Diversity, and Conflict in the Sociology of Science." *The Sociological Quarterly* 24:185–200.

Collins, Randall, and Sal Restivo. 1983b. "Robber Barons and Politicians in Mathematics: A Conflict Model of Science." *Canadian Journal of Sociology* 8:199–227.

Cozzens, Susan E., and Thomas F. Gieryn, eds. 1990. *Theories of Science in Society*. Bloomington and Indianapolis: Indiana University Press.

Crane, Diana. 1972. *Invisible Colleges*. Chicago: The University of Chicago Press.

Crozier, Michel. 1964. *The Bureaucratic Phenomenon*. Chicago: University of Chicago Press.

Dallmayr, Fred R., and Thomas A. McCarthy, eds. 1977. *Understanding and Social Inquiry*. Notre Dame: Notre Dame University Press.

Dawson, Sandra and Dorothy Wedderburn. 1980. "Introduction: Joan Woodward and the Development of Organization Theory." Pp.XIII–XL in Joan Woodward, *Industrial Organization: Theory and Practice*. 2nd ed. Oxford: Oxford University Press.

Deane, Phyllis. 1983. "The Scope and Method of Economic Science." *The Economic Journal* 93:1–12.

Denzin, Norman K. 1978. *The Research Act: A Theoretical Introduction to Sociological Methods*. 2nd ed. New York: McGraw-Hill.

Dilthey, Wilhelm. 1883. *Einleitung in die Geisteswissenschaften. Versuch einer Grundlegung für das Studium der Gesellschaft und der Geschichte*. Vol. I. Leipzig: Duncker and Humblot.

DiMaggio, Paul J., and Walter W. Powell. 1983. "The Iron Cage Revisited: Institutional Isomorphism and Collective Rationality in Organizational Fields." *American Sociological Review* 48:147–60.

Douglas, Jack D. 1970a. "Understanding Everyday Life." Pp. 3–44 in Jack D. Douglas, ed. 1970b.

Douglas, Jack D., ed. 1970b. *Understanding Everyday Life: Toward a Reconstruction of Sociological Knowledge*. Chicago: Aldine.

Douglas, Mary. 1966. *Purity and Danger: An Analysis of Concepts of Pollution and Taboo.* London: Routledge and Kegan Paul.

Douglas, Mary. 1970. *Natural Symbols.* New York: Pantheon.

Downey, H. Kirk, Don Hellriegel, and John W. Slocum Jr. 1975. "Environmental Uncertainty: The Concept and its Applications." *Administrative Science Quarterly* 20:613–29.

Dreyfus, Hubert L. 1980. "Holism and Hermeneutics." *Review of Metaphysics* 34:3–23.

Duncan, Robert B. 1972. "Characteristics of Organizational Environments and Perceived Environmental Uncertainty." *Administrative Science Quarterly* 17:313–27.

Durkheim, Emile. 1893/1965. *The Division of Labor in Society.* New York: Free Press.

Durkheim, Emile. 1895/1982. *The Rules of Sociological Method.* New York: Free Press.

Durkheim, Emile. 1912/1954. *The Elementary Forms of Religious Life.* New York: Free Press.

Edge, David O., and Michael J. Mulkay. 1976. *Astronomy Transformed: The Emergence of Radio Astronomy in Britain.* New York: Wiley.

Edmondson, Ricca. 1984. *Rhetoric in Sociology.* London: Macmillan.

Emerson, Robert M., ed. 1983. *Contemporary Field Research: A Collection of Readings.* Boston: Little, Brown and Co.

Emery, F. E., and E. L. Trist. 1965. "The Causal Texture of Organizational Environments." *Human Relations* 18:21–32.

Etzioni, Amitai. 1975. *A Comparative Analysis of Complex Organizations.* New York: Free Press.

Evans-Pritchard, Edward E. 1976. *Witchcraft, Oracles, and Magic among the Azande.* Oxford: Clarendon.

Feyerabend, Paul K. 1970. "Consolations for the Specialist." Pp. 197–230 in Imre Lakatos and Alan Musgrave, eds., *Criticism and the Growth of Knowledge.* Cambridge: Cambridge University Press.

Filstead, William J. 1970a. "Introduction." Pp.1–11 in William J. Filstead, ed., 1970b.

Filstead, William J., ed. 1970b. *Qualitative Methodology: Firsthand Involvement With the Social World.* Chicago: Markham.

Fisher, Donald. 1990. "Boundary Work and Science: The Relation between Power and Knowledge." Pp.98–119 in Susan E. Cozzens and Thomas F. Gieryn, eds., *Theories of Science in Society.* Bloomington and Indianapolis: Indiana University Press.

Fleck, Ludwik. 1935/1979. *Genesis and Development of a Scientific Fact.* Chicago: University of Chicago Press.

Freilich, Morris, ed. 1977. *Marginal Natives at Work: Anthropologists in the Field.* New York: Halstead.

236

Freudenthal, Gad. 1984. "The Role of Shared Knowledge in Science: The Failure of the Constructivist Program in the Sociology of Science." *Social Studies of Science* 14:285–95.

Fuchs, Stephan. 1986. "The Social Organization of Scientific Knowledge." *Sociological Theory* 4:126–42.

Fuchs, Stephan. 1989. "On The Microfoundations of Macrosociology: A Critique of Microsociological Reductionism." *Sociological Perspectives* 32:169–82.

Fuchs, Stephan, and Jonathan H. Turner. 1986. "What Makes a Science Mature? Patterns of Organizational Control in Scientific Production." *Sociological Theory* 4:143–50.

Fuchs, Stephan, and Mathias Wingens. 1986. "Sinnverstehen als Lebensform: Über die Möglichkeit hermeneutischer Objektivität." *Geschichte und Gesellschaft* 12:477–501.

Fuchs, Stephan, and Charles Case. 1989. "Prejudice as Lifeform." *Sociological Inquiry* 59:301–17.

Fuchs, Stephan, and Charles Case. 1990. "Toward a Grid-Group Theory of Racial Prejudice." Paper presented at the ASA Meetings in Washington, D.C..

Gadamer, Hans Georg. 1975. *Truth and Method*. New York: Seabury.

Galison, Peter. 1987. *How Experiments End*. Chicago: University of Chicago Press.

Gans, Herbert. 1982. "The Participant Observer as a Human Being: Observations on the Personal Aspects of Fieldwork." Pp.53–61 in Robert G. Burgess, ed., *Field Research: A Sourcebook and Field Manual*. London: Allen and Unwin.

Garfinkel, Harold. 1967. *Studies in Ethnomethodology*. Englewood Cliffs: Prentice-Hall.

Garfinkel, Harold, Michael Lynch, and Eric Livingston. 1981. "The Work of a Discovering Science Construed with Materials from the Optically Discovered Pulsar." *Philosophy of the Social Sciences* 11:131–58.

Geertz, Clifford. 1973. *The Interpretation of Cultures: Selected Essays*. New York: Basic Books.

Geison, Gerald L. 1981. "Scientific Change, Emerging Specialties, and Research Schools." *History of Science* 19:20–40.

Gibbs, Jack P. 1989. *Control: Sociology's Central Notion*. Urbana: University of Illinois Press.

Giddens, Anthony. 1976. *New Rules of Sociological Method*. London: Hutchinson.

Giddens, Anthony. 1982. *Profiles and Critiques in Social Theory*. Cambridge: Polity Press.

Giddens, Anthony. 1984. *The Constitution of Society: Outline of the Theory of Structuration*. Cambridge: Polity Press.

Gieryn, Thomas F. 1982. "Relativist/Constructivist Programs in the Sociology of Science: Redundance and Retreat." *Social Studies of Science* 12:279–97.

Gieryn, Thomas F., and Anne E. Figert. 1990. "Ingredients for a Theory of Science in Society." Pp. 67–97 in Susan E. Cozzens and Thomas F. Gieryn, eds., *Theories of Science in Society.* Bloomington and Indianapolis: Indiana University Press.

Gilbert, G. Nigel. 1976. "The Transformation of Research Findings into Scientific Knowledge." *Social Studies of Science* 6:281–306.

Gilbert, G. Nigel, and Steve Woolgar. 1974. "The Quantitative Study of Science: An Examination of the Literature." *Science Studies* 4:279–94.

Gilbert, G. Nigel, and Michael Mulkay. 1982. "Warranting Scientific Belief." *Social Studies of Science* 12:383–408.

Gilbert, G. Nigel, and Michael Mulkay. 1984. *Opening Pandora's Box: A Sociological Analysis of Scientists' Discourse.* Cambridge: Cambridge University Press.

Glaser, Barney, and Anselm Strauss. 1967. *The Discovery of Grounded Theory: Strategies for Qualitative Research.* Chicago: Aldine.

Glazer, Myron. 1972. *The Research Adventure: Promise and Problems of Field Work.* New York: Random House.

Goffman, Erving. 1967. *Interaction Ritual.* New York: Doubleday.

Goffman, Erving. 1974. *Frame Analysis.* New York: Harper and Row.

Goodman, Nelson. 1955. *Fact, Fiction, and Forecast.* Cambridge: Harvard University Press.

Green, Bryan S. 1988. *Literary Methods and Sociological Theory: Case Studies of Simmel and Weber.* Chicago: University of Chicago Press.

Greenwood, Ernest. 1972. "Attributes of a Profession." Pp. 3–26 in Ronald M. Pavalko, ed., *Sociological Perspectives on Occupations.* Itasca: Peacock.

Gusfield, Joseph. 1976. "The Literary Rhetoric of Science: Comedy and Pathos in Drinking Driver Research." *American Sociological Review* 41:16–34.

Habermas, Jürgen. 1962. *Strukturwandel der Öffentlichkeit.* Neuwied: Luchterhand.

Habermas, Jürgen. 1970. *Knowledge and Human Interests.* London: Heinemann.

Habermas, Jürgen. 1979. *Communication and the Evolution of Society.* Boston: Beacon Press.

Habermas, Jürgen. 1984. *The Theory of Communicative Action.* Vol. 1. Boston: Beacon Press.

Habermas, Jürgen. 1987. *The Theory of Communicative Action.* Vol 2. Boston: Beacon Press.

Habermas, Jürgen. 1990. *The Philosophical Discourse of Modernity*. Cambridge: MIT.

Hage, Jerald. 1980. *Theories of Organization*. New York: Wiley.

Hagendijk, Rob. 1990. "Structuration Theory, Constructivism, and Scientific Change." Pp. 43–66 in Susan E. Cozzens and Thomas F. Gieryn, eds., *Theories of Science in Society*. Bloomington and Indianapolis: Indiana University Press.

Hagstrom, Warren O. 1965. *The Scientific Community*. New York: Basic Books.

Hall, Richard H. 1972. "Professionalization and Bureaucratization." Pp. 276–93 in Ronald M. Pavalko, ed., *Sociological Perspectives on Occupations*. Itasca: Peacock.

Hannan, Michael T., and John H. Freeman. 1977. "The Population Ecology of Organizations." *American Journal of Sociology* 82:929–64.

Hargens, Lowell L. 1988. "Scholarly Consensus and Rejection Rates." *American Sociological Review* 53:139–51.

Harvey, Bill. 1981. "Plausibility and the Evaluation of Knowledge: A Case-Study of Experimental Quantum Mechanics." *Social Studies of Science* 11:95–130.

Hauser, Arnold. 1982. *The Sociology of Art*. London: Routledge and Kegan Paul.

Hayes, Adrian C. 1985. "Causal and Interpretive Analysis in Sociology." *Sociological Theory* 3:1–10.

Heidegger, Martin. 1962. *Being and Time*. New York: Harper and Row.

Heinz, John P., and Edward O. Laumann. 1982. *Chicago Lawyers: The Social Structure of the Bar*. New York and Chicago: Russell Sage and American Bar Foundation.

Hollis, Martin, and Steven Lukes, eds. 1982. *Rationality and Relativism*. Cambridge: MIT Press.

Hooker, Clifford A. 1987. *A Realistic Theory of Science*. Albany: State University of New York Press.

Horton, Robin. 1970. "African Traditional Thought and Modern Western Science." Pp.131–71 in Bryan Wilson, ed., *Rationality*. Oxford: Basil Blackwell.

Jagtenberg, Tom. 1983. *The Social Construction of Science: A Comparative Study of Goal Direction, Research Evolution, and Legitimation*. Dordrecht: Reidel.

Johnson, Terence. 1972. *Professions and Power*. London: Macmillan.

Kidder, Louise H., and Charles M. Judd. 1986. *Research Methods in Social Relations*. New York: Holt, Rinehart and Winston.

King, M. D. 1971. "Reason, Tradition, and the Progressiveness of Science." *History and Theory* 10:3–32.

Knorr-Cetina, Karin D. 1981. *The Manufacture of Knowledge.* Oxford: Pergamon.

Knorr-Cetina, Karin D. 1982. "Scientific Communities or Transepistemic Arenas of Research? A Critique of Quasi-Economic Models of Science." *Social Studies of Science* 12:101–30.

Knorr-Cetina, Karin D. 1983. "The Ethnographic Study of Scientific Work. Towards a Constructivist Interpretation of Science." Pp.115–40 in Karin D. Knorr-Cetina and Michael Mulkay, eds., *Science Observed.* London: Sage.

Kuhn, Thomas S. 1970. *The Structure of Scientific Revolutions.* 2nd ed. Chicago: University of Chicago Press.

Kuhn, Thomas S. 1977. *The Essential Tension: Selected Studies in Scientific Tradition and Change.* Chicago: University of Chicago Press.

Lakatos, Imre. 1970. "Falsification and the Methodology of Scientific Research Programmes." Pp.91–196 in Imre Lakatos and Alan Musgrave, eds., *Criticism and the Growth of Knowledge.* Cambridge: Cambridge University Press.

Lakatos, Imre, and Alan Musgrave, eds. 1970. *Criticism and the Growth of Knowledge.* Cambridge: Cambridge University Press.

Lankford, John. 1981. "Amateurs and Astrophysics: A Neglected Aspect in the Development of a Scientific Specialty." *Social Studies of Science* 11:275–303.

Larson, Magali. 1977. *The Rise of Professionalism: A Sociological Analysis,* Berkeley: University of California Press.

Latour, Bruno. 1980. "Is it Possible to Reconstruct the Research Process? Sociology of a Brain Peptide." Pp. 53–73 in Karin D. Knorr, Roger Krohn, and Richard Whitley, eds., *The Social Process of Scientific Investigation.* Dordrecht: Reidel.

Latour, Bruno. 1987. *Science in Action: How to Follow Scientists and Engineers through Society.* Milton Keynes, UK: Open University Press.

Latour, Bruno. 1988a. "The Politics of Explanation: An Alternative." Pp.155–176 in Steve Woolgar, ed., *Knowledge and Reflexivity.* London: Sage.

Latour, Bruno. 1988b. *The Pasteurization of France.* Cambridge: Harvard University Press.

Latour, Bruno, and Steve Woolgar. 1986. *Laboratory Life: The Construction of Scientific Knowledge.* 2nd ed. Princeton: Princeton University Press.

Laudan, Larry. 1981. "The Pseudo-Science of Science." *Philosophy of the Social Sciences* 11:173–98.

Law, John. 1986a. "Laboratories and Texts." Pp.35–50 in Michel Callon, John Law, and Arie Rip, eds., *Mapping the Dynamics of Science and Technology.* London: Macmillan.

Law, John. 1986b. "The Heterogeneity of Texts." Pp.67–83 in Michel Callon, John Law, and Arie Rip, eds., *Mapping the Dynamics of Science and Technology*. London: Macmillan.

Law, John, and R. J. Williams. 1982. "Putting Facts Together: A Study of Scientific Persuasion." *Social Studies of Science* 12:535–58.

Lawrence, Paul R., and Jay W. Lorsch. 1967. *Organization and Environment*. Cambridge: Harvard University Press.

Levitt, Barbara, and Clifford Nass. 1989. "The Lid on the Garbage-Can: Institutional Constraints on Decision-Making in the Technical Core of College-Text Publishers" *Administrative Science Quarterly* 34: 190–207.

Lincoln, James R., Mitsuyo Hannada, and Kerry McBride. 1986. "Organizational Structures in Japanese and U.S. Manufacturing." *Administrative Science Quarterly* 33: 338–64.

Lodahl, Janice B., and Gerald Gordon. 1972. "The Structure of Scientific Fields and the Functioning of University Graduate Departments." *American Sociological Review* 37:57–72.

Lofland, John. 1976. *Doing Social Life: The Qualitative Study of Human Interaction in Natural Settings*. New York: Wiley.

Luhmann, Niklas. 1976. "Generalized Media and the Problem of Contingency." Pp.507–32 in J. J. Loubser, Rainer Baum, Andrew Effrat, and Victor Lidz, eds., *Explorations in General Theory in Social Science*. New York: Free Press.

Luhmann, Niklas. 1984. *Soziale Systeme: Grundriss einer Allgemeinen Theorie*. Frankfurt/M.: Suhrkamp.

Luhmann, Niklas. 1989. *Ecological Communication*. Chicago: University of Chicago Press.

Lynch, Michael. 1985. *Art and Artifact in Laboratory Science*. London: Routledge and Kegan Paul.

Magnus, Bernd. 1988. *The Postmodern Turn: Nietzsche, Heidegger, Derrida, and Rorty*. Unpublished Manuscript: University of California/Riverside.

Manning, Peter. 1987. *Semiotics and Fieldwork*. Newbury Park: Sage.

Maranhao, T. 1986. "The Hermeneutics of Participant Observation." *Dialectical Anthropology* 10:291–309.

March, James G., and Johan P. Olsen, eds. 1979. *Ambiguity and Choice in Organizations*. Bergen, Norway: Universitetsforlaget.

Marcuse, Herbert. 1964. *One-dimensional Man*. Boston: Beacon Press.

Medawar, Peter B. 1969. *The Art of the Soluble*. London: Methuen.

Merton, Robert K. 1973. *The Sociology of Science*. Chicago: University of Chicago Press.

Merton, Robert K., and Harriet Zuckerman. 1973. "Institutionalized

Patterns of Evaluation in Science." Pp.460–96 in Robert K. Merton, *The Sociology of Science.* Chicago: University of Chicago Press.

Meyer, John W., and Brian Rowan. 1977. "Institutionalized Organizations: Formal Structure as Myth and Ceremony." *American Journal of Sociology* 83:340–63.

Meyer, John W., and W. Richard Scott, eds. 1983. *Organizational Environments: Ritual and Rationality.* Beverly Hills: Sage.

Moore, Wilbert E. 1970. *The Professions: Rules and Roles.* New York: Russell Sage.

Morgan, Gareth, ed. 1983. *Beyond Method: Strategies for Social Research.* Beverly Hills: Sage.

Morrell, Jack, and Arnold Thackray. 1977. *Gentlemen of Science.* Oxford: Clarendon.

Mulkay, Michael. 1975. "Three Models of Scientific Development." *The Sociological Review* 23:509–26.

Mulkay, Michael. 1979. *Science and the Sociology of Knowledge.* London: Allen and Unwin.

Mulkay, Michael. 1981. "Action and Belief or Scientific Discourse? A Possible Way of Ending Intellectual Vassalage in Social Studies of Science." *Philosophy of the Social Sciences* 11:163–71.

Mulkay, Michael. 1985. *The Word and the World.* London: Allen and Unwin.

Mulkay, Michael, and G. Nigel Gilbert. 1982. "Joking Apart: Some Recommendations Concerning the Analysis of Scientific Culture." *Social Studies of Science* 12:585–613.

Mullins, Nicholas. 1973. *Theories and Theory Groups in Contemporary American Sociology.* New York: Harper and Row.

Myers, Greg. 1985. "Texts as Knowledge Claims: The Social Construction of Two Biology Articles." *Social Studies of Science* 15:593–630.

Neil, Cecily C., and William E. Snizek. 1987. "Work Values, Job Characteristics, and Gender." *Sociological Perspectives* 30:245–65.

Nickles, Thomas, ed. 1980. *Scientific Discovery: Logic and Rationality.* Dordrecht: Reidel.

Perrow, Charles. 1967. "A Framework for the Comparative Analysis of Organizations." *American Sociological Review* 32:194–208.

Perrow, Charles. 1972. *Complex Organizations: A Critical Essay.* Glenview: Scott, Foresman and Co.

Perrow, Charles. 1984. *Normal Accidents: Living with High-Risk Technologies.* New York: Basic Books.

Pfeffer, Jeffrey, and Gerald R. Salancik. 1978. *The External Control of Organizations: A Resource Dependence Perspective.* New York: Harper and Row.

242   REFERENCES

Phillips, Derek L. 1977. *Wittgenstein and Scientific Knowledge.* Totowa: Rowman and Littlefield.

Pickering, Andrew. 1981. "Constraints on Controversy: The Case of the Magnetic Monopole." *Social Studies of Science* 11:63–93.

Pinch, Trevor J. 1981. "The Sun-Set: The Presentation of Certainty in Scientific Life." *Social Studies of Science* 11:131–58.

Pinch, Trevor J., and Harry M. Collins. 1984. "Private Science and Public Knowledge: The Committee for the Scientific Investigation of the Claims of the Paranormal and its Use of the Literature." *Social Studies of Science* 14:521–46.

Pinch, Trevor J., and Wiebe E. Bijker. 1987. "The Social Construction of Facts and Artifacts, Or How the Sociology of Science and the Sociology of Technology Might Benefit Each Other." Pp.17–50 in Wiebe E. Bijker, Hughes, Thomas P., and Trevor Pinch (eds). *The Social Construction of Technological Systems,* Cambridge: MIT Press.

Popper, Karl R. 1957. *The Open Society and Its Enemies.* 3rd ed. London: Routledge and Kegan Paul.

Popper, Karl R. 1961. *The Logic of Scientific Discovery.* New York: Basic Books.

Price, Derek de Solla. 1986. *Little Science, Big Science . . . and Beyond.* New York: Columbia University Press.

Pugh, Derek S., David J. Hickson, C. R. Hinings, and C. Turner. 1968. "Dimensions of Organizational Structure." *Administrative Science Quarterly* 13:65–105.

Pugh, Derek S., David J. Hickson, C. R. Hinings, and C. Turner. 1969. "The Context of Organizational Structures." *Administrative Science Quarterly* 14:91–114.

Quine, Willard V. O. 1953. *From A Logical Point of View.* Cambridge: Harvard University Press.

Ravetz, Jerome R. 1971. *Scientific Knowledge and its Social Problems.* Oxford: Clarendon.

Restivo, Sal. 1983. "The Myth of the Kuhnian Revolution." Pp. 293–305 in Randall Collins, ed., *Sociological Theory 1983.* San Francisco: Jossey-Bass.

Restivo, Sal. 1988. "Modern Science as a Social Problem." *Social Problems* 35:206–25.

Restivo, Sal. 1989. "In the Clutches of Daedalus: Science, Society and Progress." Pp.145–76 in Steven L. Goldman, ed., *Science, Technology, and Social Progress.* Bethlehem: Lehigh University Press.

Restivo, Sal. 1990. "The Social Roots of Pure Mathematics." Pp. 120–43 in Susan E. Cozzens and Thomas F. Gieryn, eds., *Theories of Science in Society.* Bloomington and Indianapolis: Indiana University Press.

Restivo, Sal, and Randall Collins. 1982. "Mathematics and Civilization." *The Centennial Review* 26:277–301.

Ritzer, George. 1980. *Sociology: A Multiple Paradigm Science.* Boston: Allyn and Bacon.

Rorty, Richard. 1979. *Philosophy and the Mirror of Nature.* Princeton: Princeton University Press.

Rorty, Richard. 1989. *Contingency, Irony, and Solidarity.* Cambridge: Cambridge University Press.

Rossi, Peter H., James D. Wright, and Andy B. Anderson, eds. 1983. *Handbook of Survey Research.* Orlando: Academic Press.

Rothenberg, Marc. 1981. "Organization and Control: Professionals and Amateurs in American Astronomy, 1899–1918." *Social Studies of Science* 11:205–25.

Rowan, Brian. 1982. "Organizational Structure and the Institutional Environment: The Case of Public Schools." *Administrative Science Quarterly* 27:259–79.

Rueschemeyer, Dietrich. 1972. "Doctors and Lawyers: A Comment on the Theory of the Professions." Pp.26–38 in Ronald M. Pavalko, ed., *Sociological Perspectives on Occupations.* Itasca: Peacock.

deSaussure, Ferdinand. 1966. *Course in General Linguistics.* New York: McGraw-Hill.

Schatzman, Leo, and Anselm Strauss. 1973. *Field Research: Strategies for a Natural Sociology.* Englewood Cliffs: Prentice-Hall.

Schutz, Alfred. 1967. *The Phenomenology of the Social World.* Evanston: Northwestern University Press.

Schwartz, Howard, and Jerry Jacobs. 1979. *Qualitative Sociology: A Method for the Madness.* New York: Free Press.

Scott, W. Richard. 1981. *Organizations: Rational, Natural, and Open Systems.* Englewood Cliffs: Prentice-Hall.

Sellars, Wilfrid. 1963. *Science, Perception and Reality.* London: Routledge and Kegan Paul.

Shapin, Steve, and Simon Schaffer. 1985. *Leviathan and the Air-Pump.* Princeton: Princeton University Press.

Shrum, Wesley, and Joan Morris. 1990. "Organizational Constructs for the Assembly of Technological Knowledge." Pp. 235–57 in Cozzens and Gieryn, eds., *Theories of Science in Society.*

Sica, Alan. 1981. "Hermeneutics and Social Theory: The Contemporary Conversation." *Current Perspectives in Social Theory* 2:39–54.

Simmel, Georg. 1890. *Über soziale Differenzierung.* Leipzig: Duncker and Humblot.

Simon, Herbert A. 1945/1976. *Administrative Behavior: A Study of Decision-Making Processes in Administrative Organization.* New York: Free Press.

Skjervheim, Hans. 1974. "Objectivism and the Study of Man." *Inquiry* 17:213–39.

Spencer, Herbert. 1898. *Principles of Sociology.* New York: Appleton.

Star, Susan L. 1983. "Simplification in Scientific Work: An Example from Neuroscience Research." *Social Studies of Science* 13:205–28.

Star, Susan L. 1985. "Scientific Work and Uncertainty." *Social Studies of Science* 15:391–427.

Stinchcombe, Arthur L. 1990. *Information and Organization.* Berkeley: University of California Press.

Strauss, Anselm. 1987. *Qualitative Analysis for Social Scientists.* Cambridge: Cambridge University Press.

Taylor, Charles. 1971. "Interpretation and the Sciences of Man." *Review of Metaphysics* 25:1–51.

Thompson, James D. 1967. *Organizations in Action.* New York: McGraw-Hill.

Tolbert, Pamela S., and Lynne G. Zucker. 1983. "Institutional Sources of Changes in the Formal Structure of Organizations: The Diffusion of Civil Service Reforms, 1880–1935," *Administrative Science Quarterly* 28:22–39.

Tosi, Henry, Ramon Aldag, and Ronald Storey. 1973. "On the Measurement of the Environment: An Assessment of the Lawrence and Lorsch Environmental Uncertainty Subscale." *Administrative Science Quarterly* 18:27–36.

Travis, Gordon L. 1981. "Replicating Replication? Aspects of the Social Construction of Learning in Planarian Worms." *Social Studies of Science* 11:11–32.

Traweek, Sharon. 1988. *Beamtimes and Lifetimes: The World of High Energy Physicists.* Cambridge: Harvard University Press.

Turner, Jonathan H., and Stephen P. Turner. 1990. *The Impossible Science: An Institutional History of American Sociology.* Beverly Hills: Sage.

Wagner, David. 1984. *The Growth of Sociological Theories.* Beverly Hills: Sage.

Warwick, Donald P., and Charles A. Lininger. 1975. *The Sample Survey: Theory and Practice.* New York: McGraw-Hill.

Wax, Rosalie. 1971. *Doing Field Work: Warnings and Advice.* Chicago: University of Chicago Press.

Weber, Max. 1922/1947. *The Theory of Social and Economic Organization.* New York: Oxford University Press.

Weber, Max. 1949. *The Methodology of the Social Sciences.* New York: Free Press.

Webster, Charles. 1982. *From Paracelsus to Newton: Magic and the Making of Modern Science.* Cambridge: Cambridge University Press.

Weick, Karl. 1979. *The Social Psychology of Organizing.* 2nd ed. Reading: Addison-Wesley.

Wellman, Barry, and S. D. Berkowitz. 1988. *Social Structures: A Network Approach.* Cambridge: Cambridge University Press.

White, Clavis J., and Peter J. Burke. 1987. "Ethnic Role Identity Among Black and White College Students." *Sociological Perspectives* 30:310–21.

Whitley, Richard. 1984. *The Intellectual and Social Organization of the Sciences.* Oxford: Clarendon Press.

Williamson, Oliver E. 1975. *Markets and Hierarchies: Analysis and Antitrust Implications.* New York: Free Press.

Wilson, Bryan R., ed. 1970. *Rationality.* Oxford: Blackwell.

Winch, Peter. 1958. *The Idea of a Social Science.* London: Routledge and Kegan Paul.

Winch, Peter. 1964. "Understanding a Primitive Society." *American Philosophical Quarterly* 1:307–24.

Wittgenstein, Ludwig. 1964. *Remarks on the Foundations of Mathematics.* Oxford: Blackwell.

Wittgenstein, Ludwig. 1967. *Philosophical Investigations.* Oxford: Blackwell.

Wolff, Janet. 1981. *The Social Production of Art.* New York: St. Martin's Press.

Woodward, Joan. 1965/1980. *Industrial Organization: Theory and Practice.* Oxford: Oxford University Press.

Woolgar, Steve. 1980. "Discovery. Logic and Sequence in a Scientific Text." Pp.239–68 in Karin D. Knorr, Roger Krohn, and Richard Whitley, eds., *The Social Process of Scientific Investigation.* Dordrecht: Reidel.

Woolgar, Steve. 1982. "Laboratory Studies: A Comment on the State of the Art." *Social Studies of Science* 12:481–498.

Woolgar, Steve. 1983. "Irony in the Social Study of Science." Pp. 239–66 in Karin D. Knorr-Cetina and Michael Mulkay, eds., *Science Observed.* London: Sage.

Woolgar, Steve. 1988a. *Science: The Very Idea.* Chichester: Ellis Horwood.

Woolgar, Steve, ed. 1988b. *Knowledge and Reflexivity: New Frontiers in the Sociology of Knowledge.* London: Sage.

Woolgar, Steve. 1988c. "Reflexivity is the Ethnographer of the Text." Pp.14–36 in Steve Woolgar, ed., *Knowledge and Reflexivity.* London: Sage.

Zuckerman, Harriet. 1977. *Scientific Elite: Nobel Laureates in the United States.* New York: Free Press.

Zuckerman, Harriet. 1988. "The Sociology of Science." Pp. 511–74 in Neil J. Smelser, ed., *Handbook of Sociology.* Newbury Park: Sage.

NAME INDEX

Abel, Theodore, 198
Adorno, Theodor W., 18
Agger, Ben, 168, 213
Aldrich, Howard, 125
Alexander, Jeffrey, 91, 100, 212f.
Apel, Karl-Otto, 194
Ashmore, Malcolm, 11, 28–30, 78, 88, 102, 212

Barnes, Barry, 19
Becker, Howard, 148–150, 165, 200, 209
Bergesen, Albert, 94
Berkowitz, S. D., 89
Bernstein, Basil, 94f.
Bazerman, Charles, 58
Ben-David, Joseph, 106
Bijker, Wiebe, 57
Blau, Peter, 113
Bloor, David, 2, 19, 24f., 34–36, 39f., 42f., 77, 94, 96, 103, 196
Bloor, M., 200, 209
Blumer, Herbert, 197, 199
Bogdan, Robert, 200
Bourdieu, Pierre, 26
Brown, James, 35
Brown, Richard, 168f., 213
Bruyn, Severine, 209
Bryman, Alan, 197
Burgess, Robert, 200, 205
Burns, Tom, 113, 116

Calhoun, Craig, 183
Callon, Michel, 49, 57, 74
Case, Charles, 94, 96f.

Chubin, Daryl, 197
Cicourel, Aaron, 70, 200
Cole, Jonathan, 144, 149, 163, 181
Cole, Stephen, 144, 149, 163, 181
Collins, Harry, 27, 29f., 53–55, 57, 64, 66, 103, 182
Collins, Randall, 7, 55, 71–73, 78, 85f., 92, 94–96, 101, 103, 106, 125f., 144f., 164, 178, 186
Crane, Diana, 163
Crozier, Michel, 114, 127

Dallmayr, Fred, 193
Dawson, Sandra, 121
Denzin, Norman, 202–204
Dilthey, Wilhelm, 194
DiMaggio, Paul, 135
Douglas, Jack, 198
Douglas, Mary, 94
Dreyfus, Hubert, 200
Durkheim, Emile, 16, 36f., 39f., 63, 77, 93f., 186, 198

Edge, David, 106
Edmondson, Ricca, 168f.
Etzioni, Amitai, 125f., 129
Evans-Pritchard, Edward, 23, 25
Emerson, Robert, 200

Feyerabend, Paul, 20
Filstead, William, 199, 201
Fisher, Donald, 17
Fleck, Ludwik, 3
Foucault, Michel, 18

247

SUBJECT INDEX

Accumulative advantage, principle of, 72f.

Art (as profession), 145–165, 172; see also Literature, Professions

Behaviorism, 197f., 206

Bureaucracy, 113, 116, 191f.; see also Organizational theory, Organizations

Classics (in sociology), 63, 91, 213; see also Social theory, Sociology of sociology

Cognitive styles
   in mundane groups, 83, 93–97, 109, 155f.
   in organizations, 87f., 140
   in science, 7–9, 16, 81, 83, 91f., 97–102, 109, 140, 177–192

Competition, 86, 103f., 105–108, 134f., 155, 163, 180, 186–189, 214; see also Scientific change

Constructivism, 14f., 78; see also Sociology of scientific knowledge
   interpretive, 3f., 6, 8, 10, 102
   structural, 6f., 77f.

Context of discovery, 2, 35

Context of justification, 2, 35, 42

Controversies; see Scientific controversies

Conversation, 7, 33, 58f., 78, 90, 92, 96f., 102, 165, 167, 171f., 174, 167, 212ff.; see also Fragmentation, Hermeneutics, Textuality

Coupling; see Social density

Critical theory, 39, 100

Cumulation; see Scientific change

Deconstructionism, 9f., 11–14, 17f., 175, 196

Discourse analysis, 20, 31, 58–61, 66, 88, 170, 197

Discoveries, 4, 28, 68, 84, 187f.; see also Innovation

Double hermeneutic, 194

Durkheimian theory; see also Neo-Durkheimian theory
   of knowledge, 6, 36–42, 77, 93f.
   of religion, 21–23, 36, 40, 77

Empirical program of relativism (EPOR), 20, 29f., 53f., 103, 108

Epistemology; see Philosophy of science

Ethnomethodology; see sociology of scientific knowledge (and ethnomethodology)

Ethos of science; see Scientific organizations (ethos and norms)

Facts
   as constructions, 14, 27–31, 33, 48–76, 79, 84, 88f., 144
   as mirrors, 20, 28, 47, 91, 168
   as statements, 46–48, 55, 71, 88f.
   as black boxes, 47, 50–52, 56, 62, 64f., 69, 75, 77, 88f., 162, 167, 171f.

Fragmentation; see Scientific change

Fraud, 42

251

252   SUBJECT INDEX

Garbage-can theory; *see* Organiza-
tional theory
Grid/group theory, 93ff., 185; *see also*
NeoDurkheimian theory
Grounded theory, 199, 205, 210

Hermeneutics, 9, 16, 25, 90, 100,
193–215; *see also* Interpretive
paradigm, Conversation

Incommensurability, 26f., 54
Innovation, 16, 92, 107, 141, 143,
148–175, 185, 187, 190; *see also*
Discoveries
Institutionalism; *see* Organizational
theory
Interpretive paradigm, 3–7, 9, 16, 20,
27–31, 183, 193–206
Invisible colleges; *see* Research fronts

Laboratories, 5, 78, 112, 173
as lifeworlds, 66, 83f.
as supporting agents, 33, 57f., 67–
71, 74f., 180f.
as workplaces, 19f., 66, 143f., 162,
173
Law (as profession), 145–160; *see*
also Professions
Lay interpretations, 147, 160f., 173f.,
183f., 194, 196ff., 202ff., 206,
208
Literary criticism, 9–12, 14, 32–34,
166–175, 196
Literature (as profession), 7, 15, 98,
165–175; *see also* Professions

Matthew effect, 75
Medicine (as profession), 9, 145–160;
see also Professions
Member validation, 183, 200, 209
Mertonian paradigm; *see* Sociology of
science
Methodenstreit, 193–196, 212

Migration; *see* Scientific change
Mutual dependence, 81f., 85–108,
114, 117–129, 136–142, 147,
154–175, 178–180, 184–189,
192, 210ff.; *see also* Social
density

NeoDurkheimian theory, 7, 15, 77f.,
93–109, 155, 159, 175, 177–
192; *see also* Durkheimian theory, Grid/group theory
Networks of support, 4, 10f., 48–53,
55, 64f., 171f.
laboratory hardware, 67–70, 75,
88, 97
numbers and statistics, 57, 63f., 75,
88
organizations, 55, 73–75, 161
property, 71–75
references and other people, 48,
63–66, 75, 88, 161, 167
textuality, 57–62, 75
Network theory, 10, 14, 20, 57–76,
89
New literary forms, 11, 13f., 30, 34
Normal accidents, 102–108
Normal science, 8, 19, 56, 66, 85,
105f., 149, 187, 202

Organizational theory, 17, 129ff., 177
classical/neoclassical, 87f., 111,
121, 129–132, 135–142
control, 125–129, 145
garbage can, 87, 129, 132f., 135–
142
institutionalist, 4, 87, 129, 133–
142, 147
technological, 7, 15, 104f., 111–
142, 178, 191f.
Organizations; *see also* Scientific
organizations
control, 15f., 81f., 112–121, 125–
129, 141, 144, 208–214
environments, 112, 121–125,
128f., 131, 134f., 143